AF762968

Creative Industries in Europe

Caroline Chapain • Tadeusz Stryjakiewicz
Editors

Creative Industries in Europe

Drivers of New Sectoral and Spatial Dynamics

Springer

Editors
Caroline Chapain
Business School
University of Birmingham
Birmingham, UK

Tadeusz Stryjakiewicz
Institute of Socio-Economic Geography and Spatial Management
Adam Mickiewicz University
Poznań, Poland

ISBN 978-3-319-56495-1 ISBN 978-3-319-56497-5 (eBook)
DOI 10.1007/978-3-319-56497-5

Library of Congress Control Number: 2017942812

© Springer International Publishing AG 2017
This work is subject to copyright. All rights are reserved by the Publisher, whether the whole or part of the material is concerned, specifically the rights of translation, reprinting, reuse of illustrations, recitation, broadcasting, reproduction on microfilms or in any other physical way, and transmission or information storage and retrieval, electronic adaptation, computer software, or by similar or dissimilar methodology now known or hereafter developed.
The use of general descriptive names, registered names, trademarks, service marks, etc. in this publication does not imply, even in the absence of a specific statement, that such names are exempt from the relevant protective laws and regulations and therefore free for general use.
The publisher, the authors and the editors are safe to assume that the advice and information in this book are believed to be true and accurate at the date of publication. Neither the publisher nor the authors or the editors give a warranty, express or implied, with respect to the material contained herein or for any errors or omissions that may have been made. The publisher remains neutral with regard to jurisdictional claims in published maps and institutional affiliations.

Printed on acid-free paper

This Springer imprint is published by Springer Nature
The registered company is Springer International Publishing AG
The registered company address is: Gewerbestrasse 11, 6330 Cham, Switzerland

Contents

Chapter 1
Introduction – Creative Industries in Europe: Drivers of (New) Sectoral and Spatial Dynamics

Caroline Chapain and Tadeusz Stryjakiewicz

1.1 Creative Industries in Europe – Rise of a Discourse

In the last 15 years, more and more attention has been given to the creative industries as a new motor of economic growth worldwide (UN 2008, 2010, 2013; EY 2015). Started in Australia in 1994 and then embraced in the UK in 1998, this trend has slowly spread across Europe, supported by European policy discourses, mapping exercises and initiatives to support these industries. Since the mid-2000s, European institutions have started associating the word 'creative' to the then so called cultural sectors – which became the cultural and creative sectors – giving more focus to their direct economic contribution (CEU 2007). This shift was associated with the production of increasing economic evidence either by the European Commission itself or consultants and private actors (KEA 2006; ERICARTS 2010; Power 2011; EY 2014; TERA 2014; EC 2016) and the adoption of policy initiatives to support the cultural and creative industries (CCI) through mainstream EU support programmes (EP 2013) such as the structural fund (EU 2012) or specific ones such as Creative Europe (EU 2013) or the European Creative Industries Alliance (ECIA 2014). Today, the cultural and creative industries are seen as an important contributor to the European innovation dynamic, economic growth and overall development, participating in its industrial renaissance (EC 2014) and as such should be supported both at the European level but also locally, regionally and nationally (EP 2013; ECIA 2014; EC 2016).

C. Chapain (✉)
Business School, University of Birmingham, Birmingham, UK
e-mail: c.a.chapain@bham.ac.uk

T. Stryjakiewicz
Institute of Socio-Economic Geography and Spatial Management, Adam Mickiewicz University, Poznań, Poland
e-mail: tadek@amu.edu.pl

© Springer International Publishing AG 2017

C. Chapain, T. Stryjakiewicz (eds.), *Creative Industries in Europe*,
DOI 10.1007/978-3-319-56497-5_1

It is important to note that this recognition has taken some time and has not been exempt of debate in terms of definitions, measurement and the overall role of CCI in both economy and society and how best to support them. In 1999, echoing the growing interest in the contribution of culture and cultural industries to economic development in Europe (Bianchini and Parkinson 1993), the European Parliament published a working paper recognising the potential of activities within the cultural industries to support innovative activities and job creation but also European identity within the Europe of Fifteen (EP 1999). As such, prospect of growth were seen in cultural tourism and education. The paper also highlighted the lack of definitional clarity with regards to what constitutes the cultural sector.

Despite this early publication, it was not until the mid-2000s that the evidence and rhetoric around the economic contribution of the cultural and creative industries and how to support them started gaining momentum at the European level. Indeed, the study on the economy of culture by KEA (2006) sponsored by the European Commission marked a key step by providing the first methodological discussion and detailed statistics on '*the direct and indirect socio-economic impact of the cultural sector in Europe, thereby assessing its contribution to the Lisbon agenda, and in particular in terms of realising growth, competitiveness, more and better jobs, sustainable development, and innovation*' (ibid.: 1). The study reflected on the difficulties to create a European database to evidence this contribution and advocated for some European standards. Since then, EUROSTAT has launched a compendium of European cultural statistics based on a common framework: the first one was in 2007, followed up by two updates in 2011 and 2016 (EUROSTAT 2007, 2011, 2016).

From a definitional point of view, the 2006 KEA report posed the base for the early delineation of the cultural and creative sectors within the European Union. Contrary to the 1998 UK definition which amalgamated 13 sectors together, i.e. 'Advertising, Architecture, Arts and antique markets, Computer and video games, Crafts, Design, Designer Fashion, Film and video, Music, Performing arts, Publishing, Software and Television and Radio' under the new banner of the creative industries (DCMS 1998), KEA (2006) established some distinctions between core art fields (visual arts, performing arts and heritage), cultural industries (film and video, television and radio, video games and music), creative industries (design, architecture and advertising) and related industries (such as manufacturing sector of products related to cultural and creative goods, i.e. PC and MP3 players, ICT). These distinctions are in line with other academic works which have attempted to categorise cultural and/or creative activities with regards to their position along the cultural and creative value chain, i.e. if they are focusing on producing cultural and creative content or only reproducing and disseminating this content or with regards to the type of outputs these activities produce, i.e. experience, services, original/content (see Hesmondalgh 2007; Throsby 2008; O'Connor 2010 for a review). The distinction between cultural and creative industries has been an important feature of the European discourse as illustrated by the fact that most official documents use the concept 'cultural and creative sectors' (CEU 2007, 2015; EC 2010a, b, 2011, 2012, 2016; EP 2013; OMC-EWG 2010,

2014). This distinction is summarised as follows by the European Commission (EC 2010a: 5): '*"Cultural industries" are those industries producing and distributing goods or services which at the time they are developed are considered to have specific attribute, use or purpose which embodies or conveys cultural expressions, irrespective of the commercial value they may have. (...) "Creative industries" are those industries which use culture as an input and have a cultural dimension, although their outputs are mainly functional.*'

This distinction emphasises the non-economic role of culture and cultural sectors within society and has permeated the European discourse in the last 10 years; it relates to the European Union promotion of cultural and social development, such as cultural diversity and building a European identity as well as supporting social cohesion among European countries (CEU 2010; EP 2013). This is in line with an increasing focus on understanding the role of culture-based creativity and on capturing both the direct and indirect impacts of the cultural and creative industries overall (KEA 2009; EU 2012; Flemming 2015). This is part of a long term debate on the direct and indirect contribution of the arts and culture to the economy and society in terms of economic outcomes (such as employment, value added, attractivity, tourism, innovation), social and cultural outcomes (such as cultural and social capital, wellbeing, social cohesion, skills) and environmental outcomes (regeneration of places). This debate has been renewed by the ongoing discussion around the transformational role that creative occupations and industries play within the wider economy in terms of innovation and spillovers today (see Chapain and Hargreaves 2016 for a review). This on-going discussion has led to some argument on the best way to measure the CCI from either an industrial and/or an occupational perspective. This has been illustrated by recent changes in the measurement and definition of CCI in the UK since 2010.

Indeed, in 2010, after 12 years of focusing on a definition of the creative industries based on the 13 sectors mentioned above (DCMS 1998), the UK government decided to revise its approach both in terms of the sectors included (museums, galleries, archives and libraries until then excluded) and in extending its measurement to creative occupations embedded within the non-creative industries (DCMS 2010). This corresponds to a shift to consider a wider view of the creative economy as advocated by Bakshi et al. (2013) in order to better take into account the direct and indirect economic impacts of both the cultural and creative sectors and culture-based creativity within the rest of the economy through creative occupations. In a way, this approach which encompasses both industries and occupations links to the creative class rhetoric developed by Richard Florida (2002) about the role that occupations categorised as creative can play in supporting economic growth today. Florida's discourse has certainly played a role in the development of the creative rhetoric in Europe in the last 15 years (Florida and Tinagli 2004; Musterd and Murie 2010; Musterd and Kovacs 2013). While most of the empirical evidence developed at the European level has measured and captured the economic impacts of the creative industries using an industrial approach as illustrated by KEA (2006), TERA (2010, 2014), Power (2011), Ketels and Protsiv (2014) and EY (2014), the publications by EUROSTAT (2007, 2011, 2011) offer data on both industries and occupations, and the European Competitiveness Report 2010 (EC 2010b) offered some data associated with the wider creative class as well.

The industrial approach has nevertheless tended to prevail within official documents published by the European Commission and the Parliament. However, reflecting the ever evolving debate on the role of the creative industries within the economy and society and the attempt to better measure their various impacts, the list of creative sectors included in European reports has slightly varied depending on documents – see Table 1.1. While there has been a relative consistency along the cultural and creative sectors identified by KEA (2006), some reports have adopted narrower views depending on their focus and, more recently, there has been a tendency to include some related manufacturing activities incorporating high creative content such as high-end industries (Frontiers economics 2012, EC 2016) to account for the indirect contribution of the CCI further down the value chain. It is important to note though that while the European Union has developed various definitions, reports and statistics to understand the sector, it has recognised early on the wide diversity of terms associated with both cultural and creative industries across Europe (EC 2010a) and has emphasised the need to consider the particularities of local and national creative ecosystems (EP 2013). Nevertheless, most of the sectoral evidence developed so far offer overall European analysis with some cross-sectoral and/or national comparisons as discussed in the next section.

1.2 Creative Industries in Europe: Evidence on the Spatial and Sectoral Dynamics So Far

The latest reports available on the sectoral dynamics of the cultural and creative industries at the European level are the latest EUROSTAT *Cultural statistics* report (2016) and the European Commission (2016) report on *Boosting the competitiveness of the cultural and creative industries for growth and jobs*. In this section, we discuss some of the findings of the EUROSTAT (2016) report. While being a bit more conservative in its definition of the cultural and creative industries sector (see Table 1.1), it offers a detailed examination of cross-national sectoral dynamics.

Combining an industrial and occupational approach, EUROSTAT (2016) estimated that 6.3 million people were working in a cultural occupation or cultural sector in 2014, i.e. 2.9% of all EU employment. This proportion varies across EU countries from 1.1% in Romania to 5.2% in Luxembourg. Overall, the employment in the cultural sectors has been more resilient than the rest of the economy during the 2008–2009 crisis, demonstrating an increase in jobs of 0.7% while EU employment was decreasing by 1.4% from 2008 to 2010. Similarly, cultural jobs have increased more rapidly than the rest of the European economy from 2011 to 2014 (1.3% versus 0.3%). This overall trend hides important variations across EU countries though. For example, the cultural sectors were more affected than the rest of the national economies in Bulgaria, Greece, Hungary, Lithuania, Malta,

Table 1.1 Cultural and creative industries definitions – key European documents and reports 2006–2016

Document/Report	Sectors included
KEA (2006) – The economy of culture in Europe	Arts – visual arts, performing arts and heritage
	Cultural industries – film and video, television and radio, video games, music, books and press
	Creative industries – design (fashion design, graphic design, interior design and product design), architecture and advertising
EUROSTAT (2007) – Cultural statistics	Visual arts, performing arts and heritage, film and video, television and radio, video games, music, books and press (including manufacturing and retail parts), architecture
European Commission (2010a) – Green paper: unlocking the potential of cultural and creative industries	As per KEA (2006) report – see above
TERA (2010) – Building the digital economy	Core creative industries: press and literature, music, video, software and database
	Non-core industries: a) Interdependent industries: TV, radio, CD and DVD players, computers and equipment, musical instruments, photographic and cinematographic instruments, blank recording material, paper;
	b) Non-dedicated support industries: wholesale and retail, transport and telephony and Internet
European Commission (2010b) – European competitiveness report 2010	Information services: books and publishing, music, film, video, TV and radio, software and computer programming
	Business services: architecture, advertising, design, photography
	Arts and entertainment: creative arts and entertainment activities
EUROSTAT (2011) – Cultural statistics	Same as EUROSTAT (2007) taking into account changes in industrial classification
Power (2011) – Priority report: creative and cultural industries	Books and publishing, music, video, film, TV and radio, software and computer programming, advertising, photography, museums, galleries and libraries, visual arts, performing arts
TERA (2014) – The contribution of the creative industries to EU GDP and employment evolution 2008–2011	Core creative industries: film and video, video games, broadcasting, music, books and press publishing
	Non-core creative industries:
	a) Interdependent industries: They include for instance the manufacturing, wholesale and retail sale of television sets, radios, CD

(continued)

Table 1.1 (continued)

Document/Report	Sectors included
	players, DVD players, electronic games equipment, computers, musical instruments, blank recording material, paper, photocopiers, and photographic and cinematographic instruments;
	b) Non-dedicated support industries: wholesale and retail, transport and telephony and Internet
Ketels and Protsiv (2014) – European cluster panorama 2014	Creative industries: music, video, communication, performing arts, marketing and printing
EY (2014) Creating growth. Measuring cultural and creative markets in the EU	Cultural and creative industries: books, newspapers and magazines, music, performing arts, TV, film, radio, video games, visual arts, architecture, advertising
EUROSTAT (2016) Cultural statistics	Cultural domains: heritage, archives, libraries, book and press, visual arts (plastic arts, photography, design), performing arts (music, dance, drama, combined arts and other live shows), audiovisual and multimedia (film, radio, TV, video, sound recordings, multimedia works, video games), architecture, advertising and art crafts
European Commission (2016) Boosting the competitiveness of the cultural and creative industries for growth and jobs	Cultural and creative industries:
	Core CCIs: advertising, architecture, archives, libraries, cultural heritage, book and press, cultural education, design (incl. fashion design), visual arts, music, performing arts and artistic creation, radio and TV, software and games, video and film
	Fashion industry: manufacturing of intermediate goods and fashion good, wholesale and retail
	High end industries: high end fashion, jewellery and watches, accessories, leather goods, perfume and cosmetics, furniture and households appliances, cars, boats, gastronomy, hotels and leisure, retail and auction houses, publishing

Romania, and Spain, during the period 2008–2010. However, these trends reversed during the period 2011–2014 with the CCI growing above national averages in these countries like in most European countries except Germany, Estonia, France and Finland. Data in terms of numbers of European cultural enterprises show that they accounted for 6.3% of all firms in services in 2013: this proportion varies from 3.2% in Lithuania to 8.4% in Belgium and Sweden. These cultural companies contributed to 5.3% of the EU turnover in services; again this proportion varied from 2.2% in Cyprus and Latvia to 6.4% in the UK. Most of the value added in the European cultural sectors is generated by the publishing, film and TV sectors. Small and medium-sized enterprises tend to dominate the cultural sectors even though they tend to be less productive than large cultural companies (EUROSTAT 2016). Finally data on extra- EU cultural trade show that while the EU was experiencing a negative balance of trade in terms of cultural goods in 2008, this has shifted to a surplus in 2013. This is due to a growth in exports and a decrease in imports. Trends again vary by countries. Cultural goods represent less than 1% of all EU exports; this proportion raises to 2% of all UK exports. EU trade is split between intra EU and extra EU with a slight advantage to intra-EU exports overall. However, some countries such as the UK, Sweden, France, Portugal and Luxembourg export in greater proportions outside of the EU whereas the great majority of countries import mostly from the EU except the Netherlands and the UK. The biggest exports of cultural goods consist in books, works of art and antiques whereas the biggest imports are for films and video games, works of art and books. Finally, the EU exported 32.9% of its cultural goods to the USA and 21% to Switzerland whereas the cultural goods it imported comes principally from China (41.1%) and the USA (27.9%) in 2014.

These findings offer interesting insights into the overall European dynamic of the cultural industries sectors as well as suggesting very different situations from one country to another in terms of the importance of the CCI within their respective national economies, their growth and trade patterns and how the 2008–2009 crisis may have affected them. While useful, this analysis does not shed much light on the wider economic and social institutional contexts within which these industries develop both at the national levels but also at the local and regional levels. While the European regulatory environment and the existence of European programmes to support the creative industries are important in supporting the development of these industries in Europe, the roles of national policies as well as local and regional environments are seen as critical in this process as highlighted by various EU documents (EC 2010a; ERICARTS 2010; CEU 2010; EU 2012; European Cluster Observatory 2013; EP 2013; Ketels and Protsiv 2014; EC 2016).

In addition to the UK, various national governments across Europe have early on adopted the creative discourse in their policies (Braun and Lavanga 2007). In its latest address on the topic, the European Parliament emphasises the role of member states in recognising the economic potential of the CCI (EP 2013). Nevertheless, cities and regions have played and are seen as playing a key role in supporting the development of CCI (Hall 2000; Florida 2002; EC 2010a; Musterd and Murie 2010; Musterd and Kovacs 2013; EC 2016). This is due to the fact that CCI tend to cluster in space (Chapain et al. 2010; Power 2011; Ketels and Protsiv 2014) to benefit from economies of scale and agglomeration, i.e. the various actors present at the different stages of the creative value chain from university graduates to creative workers, brokers, public institutions, arts galleries, theatres, creative associations... as well as from access to other local and regional economic actors. Some national CCI mapping exercises have emerged in Europe in recent years (Chapain et al. 2010;

Boix et al. 2013; Clifton et al. 2013; Trippl et al. 2013; Boix et al. 2014). While helpful, they tend to concentrate on Western Europe.

Two European reports are nevertheless worth mentioning as they attempt to offer some comparative insights in terms of the geographical distribution of the creative industries across countries from both Western and Central and Eastern Europe: the priority sector report on the cultural and creative industries by Power (2011) and the European cluster panorama of Ketels and Protsiv (2014). The work by Power (2011) covers most of the CCI as defined by the European Commission (EC 2010a). Using data on employment from 2009, the report confirmed the tendency of the CCI to concentrate in large urbanised areas with high prosperity levels across Europe particularly large city regions in Western Europe such as London, Paris, Milan, Madrid, Barcelona, Rome, Athens or Munich. Nevertheless, the report also highlighted that some Central and Eastern European cities like Bratislava, Budapest and Praha displayed important shares of their labour markets within the creative industries. In addition, the regions and areas which were experiencing the most important growth in terms of CCI tended to be small and medium-sized cities located in more peripheral areas and/or countries in Central Eastern Europe such as Latvia, Slovakia, Estonia, Romania, Lithuania, Slovenia and Bulgaria over the period 2004–2009. It is important to note though that due to deficiencies in local and regional statistics in this part of Europe, these findings must be treated with caution and would need further verifications. The more recent report by Ketels and Protsiv (2014) only covers a subset of CCI (music, video, performing arts, marketing, printing and communication). Using employment data for 2012, Ketels and Protsiv (2014) confirm the early findings of Power (2011).

These findings while confirming the overconcentration of CCI in the largest cities of Western Europe also allude to some interesting emerging spatial dynamics within Central and Eastern Europe and smaller and medium-sized cities. However, these reports only offer an aggregate perspective without detailed insights on national dynamics. While, as mentioned above, there have been, in recent years, some emerging creative industries mapping exercises undertaken for Western European countries, there is a need for more research of this type especially within Central and Eastern European contexts. In addition, as the creative industries tend to cluster in cities, especially medium and large ones, much initial research has been published analysing creative industries dynamics in such settings, leaving aside other dynamics in smaller cities and rural areas. It is important to note that much of the research on large and medium-sized cities do not take into account the impacts of national geographical creative dynamic and economic idiosyncrasies despite increased evidence to their importance (Andres and Chapain 2013; Comunian et al. 2014; Clifton et al. 2016). As highlighted in recent literature, there is still much to do to develop a truly European perspective on this discourse (Comunian et al. 2014; Clifton et al. 2016). Related to this, while much work has been published on the clustering phenomenon of the creative industries and the reasons and detailed dynamics of particular creative clusters across Western Europe, very little work has been published on similar phenomena in Central and Eastern Europe. As such, there is both a lack of national perspectives on the regional and national creative dynamics in addition to a lack of research examining clustering processes in Central and Eastern Europe preventing real comparative analysis between the various geographical European contexts. In part, this is due to a lag in the adoption of the creative rhetoric in Central and Eastern Europe, a lack of data and specialists on the topic as well as politician's concentration on the problems of traditional industries and 'non-creative' labour market. It should be emphasised that, although the scientific literature on the post-socialist transformation is extensive (Sztompka 1995; Andrusz et al. 1996; Sailer-Fliege 1999; Stenning 2004; Tsenkova and Nedović- Budić 2006; Kovács et al. 2007a; Stanilov 2007; Stryjakiewicz et al. 2010a; Węcławowicz et al. 2010; Haase et al. 2011), so far very few scholars have focused on the development and dynamics of the creative economy and creative industries (although the situation has started to change recently). Many of the new contributions result from large-scale research projects (Švob-Đokiæ 2007; Ságvári and Lengyel 2008; Klasik 2009; Cikánek 2009; Kloudouvá 2009; Rumpel et al. 2010; Stryjakiewicz 2010; Stryjakiewicz et al. 2010b, 2012, 2014, 2015; Borsekova et al. 2015; Kasprzak 2015; Nurkovic 2015; Tomczak and Stachowiak 2015) and others from research contracted by local authorities (mostly related to renewal programmes). The latter are often focused on purely local problems, have a descriptive character and produce stock-taking types of documents. These do not deal with factors and mechanisms explaining the development of the creative sectors in different geographical scales and institutional settings. There is also a scarcity of comparative and generalising studies. Combined to the lack of full geographical coverage on the distribution of the creative industries and their dynamic across Europe up to now, there is also a need to better document the impact that the 2008–2009 crisis has had on the development and distribution of the creative industries across European countries. Indeed, the overall sectoral analysis discussed above suggests that CCI have been less affected by the crisis and have since then recovered more quickly than the rest of the European economy. However, this trend has not been uniform across countries, especially in Central and Eastern Europe. More evidence is needed about these particular dynamics. This book is an attempt to address these issues in a comparative perspective.

In addition, within the wider research on the sectoral and spatial dynamics of the creative industries, some more specific elements are in need of more research. Indeed, despite the existence of numerous studies on the phenomenon of creative clustering (see for example, Cinti 2008; Evans 2009; Chapain and Comunian 2010; Chapain and Sagot-Duvauroux 2015 for a review), some aspects of creative clusters are still in need of further exploration such as their innovation performance and geographical dynamic, the question of creative specialisation or diversification or the multiscalar dimensions of creative industries location patterns in the face of the globalisation of some of these industries (Flew 2013). These are particularly relevant within the European context given the increasing focus on the innovation capacity of the creative industries (ECIA 2014; Flemming 2015), the role of clusters (ECO 2013), the potential of European creative markets, the importance of cross-regional, cross-border and international collaboration (EP 2013; EC 2016) as well as smart specialisation within the European agenda (EC 2014). Finally, more and more focus has been put on developing creative partnerships at the local and regional levels to support the CCI and to better take into account the role of the local environment within European documents (EC 2010a; OMC-EWG 2014; EP 2013). Within this policy context, it is important to note the recent call by Jakob and Van Heur (2015) for more richly situated analysis of the role of cultural and creative intermediaries, i.e. 'arts and cultural councils, policy networks, economic development agencies, foundations and unions to artist collectives, cultural centres, creative industries, incubators, festivals and tradeshows' (ibid.: 357) in supporting the creative economy at the local and regional levels.

1.3 Objectives and Structure of the Book

Taking a purposeful geographical approach to the study of the creative industries across various Western and Central and Eastern European contexts since the 2008–2009 recession, this book offers novel contributions filling some of these gaps by exploring in more depth the sectoral and spatial dynamics of the creative industries within underrepresented areas of research and geographical contexts. The book is divided in two parts focusing on particular geographical scales and creative processes.

The first part offers new insights on creative clustering processes and trends, on the one hand, and on the interplay between particular cultural intermediaries and the development of the creative sectors within particular cities, on the other hand. First, the chapter by Stachowiak and Stryjakiewicz takes a wide perspective and interrogates the processes of creative specialisation and diversification within the European context. The chapter by Molho and Sagot-Duvauroux follows by adopting a multiscalar approach to examine the location patterns of arts galleries in New York, Istanbul and Singapore demonstrating the interplay between local and global sectoral dynamics. In contrast, the chapter by Chapain and Stachowiak offers a detailed and local analysis of and original data on the innovation performance and dynamic of the Soho film cluster in London. The remaining chapters by Zajtmann and Lange then shift our focus to the role of particular cultural intermediaries in supporting the development of the creative industries at the local level. The chapter by Zajtmann discusses the role that a local professional association has had in making Paris the centre of the global fashion industry, whereas the chapter by Lange examines the roles that particular creative scenes in Berlin and Leipzig have played in supporting the development of creative entrepreneurs.

The second part offers new perspectives on the spatial distribution and economic performance of the creative industries in different national and regional European economic contexts: Spain (chapter by Pareja-Eastaway and Pradel i Miquel), Poland (chapter by Kasprzak), Romania (chapter by Becut and Craciun), Czechia (chapter by Slach and Ženka) and Slovenia (chapter by Kozina and Bole). One of the key dimensions discussed in these contributions relates to the extent to which the creative industries have offered viable opportunities for economic growth following the 2008–2009 economic crisis. In addition, these chapters offer a detailed insights into the socio-economic and institutional contexts within which the CCI have developed sectorally and spatially.

References

Andres L, Chapain C (2013) The integration of cultural and creative industries into local and regional development strategies in Birmingham and Marseille: towards an inclusive and collaborative governance? Reg Stud 47(2):161–182
Andrusz G, Harloe M, Szelényi I (eds) (1996) Cities after socialism. Blackwell, Oxford
Bakhshi H, Hargreaves I, Mateos-Garcia J (2013) A manifesto for the creative economy. NESTA, London
Bianchini F, Parkinson M (eds) (1993) Cultural policy and urban regeneration: the west European experience. Manchester University Press, Manchester
Boix R, Lazzaretti L, Capone F, De Propris L, Sánchez D (2013) The geography of creative industries in Europe: comparing France, Great Britain, Italy and Spain. In: Lazzeretti L (ed) Creative industries and innovation in Europe. Routledge, Oxon
Boix R, Hervas-Oliver J-L, De Miguel-Molina B (2014) Micro-geographies of creative industries clusters in Europe: from hot spots to assemblages. Papers in Regional Science, online first
Borseková K, Cole D, Petríková K, Vaňová A (2015) Nostalgic sentiment and cultural and creative industries in regional development: a Slovak case study. Quat Geochronol 34(2):53–63
Braun E, Lavanga M (2007) An international comparative quickscan into national policies for creative industries. Euricur, Rotterdam
Chapain C, Comunian R (2010) Enabling or inhibiting the creative economy: the role of the local and regional dimensions in England. Reg Stud 44(6):717–734
Chapain C, Hargreaves I (2016) Citizenship in the creative economy. In: Hargreaves I, Hartley J (eds) The creative citizen unbound: how social media and DIY culture contribute to democracy, communities and the creative economy. Policy Press, Bristol
Chapain C, Sagot-Duvauroux D (2015) Cultural and creative cluster, district or quarter: conceptual evolution and overview. Preliminary findings. Presented at the regional studies cultural and creative cluster seminar, Angers, France, 8–9 June
Chapain C, Cooke P, De Propris L, Mac Neill S, Mateos-Garcia J (2010) Creative clusters and innovation. Research report. NESTA, London. http://www.nesta.org.uk/publications/reports/assets/features/creative_clusters_and_innovation_report. Accessed 1 Sept 2016
Cikánek M (2009) Kreativni průmysly – přiležitost pro novou ekonomiku. Divadelni ústav, Praha
Cinti T (2008) Cultural clusters and districts: the state of the art. In: Cooke P, Lazzaretti L (eds) Creative cities, cultural clusters and local economic development. Edward Elgar, Cheltenham
Clifton N, Cooke P, Kalso HH (2013) Towards the reconciliation of the 'context-less' with the 'space-less'? The creative class across varieties of capitalism: new evidence from Sweden and the UK. Reg Stud 47(2):201–215
Clifton N, Chapain C, Comunian R (eds) (2016) Creative regions in Europe. Routledge, Oxon
Comunian R, Chapain C, Clifton N (2014) Creative industries and creative policies: a European perspective? City Cult Soc 5(2):51–53
Council of European Union [CEU] (2007) Contribution of the cultural and creative sectors to the achievement of the Lisbon objectives – adoption of the council conclusions. Introductory note. Cult 29, 9021/07, Brussels 8 May 2007
Council of European Union [CEU] (2010) Council conclusion on the contribution of the culture to local and regional development –3013th Education, Youth and Culture Council meeting, Brussels, 10 May 2010

Council of European Union [CEU] (2015) Council conclusions on cultural and creative cross-over to stimulate innovation, economic sustainability and social inclusion. Off J Eur Union, 27th May 2015, C172/13

Department for Media Culture and Sport [DCMS] (1998) Creative Industries Mapping Document. Department for Culture, Media and Sport, London

Department for Media Culture and Sport [DCMS] (2010) Creative industries economic estimates. Technical note, February. DCMS, London

Economics F (2012) The value of the cultural and creative industries to the European economy. Prepared for ECIA. Frontiers Economics, London

ERICARTS (2010) Study on the contribution of culture to local and regional development. Evidence from the structural funds. Final report. ERICARTS, Brussels.

Ernest and Young [EY] (2014) Creating growth. Measuring cultural and creative markets in the EU. Studio EY, Cannes

Ernest and Young [EY] (2015) Cultural times. The first global map of cultural and creative industries. Ernest and Young. http://www.worldcreative.org/. Accessed 1 Sept 2016

European Cluster Observatory [ECO] (2013) Creative industries Analysis of industry-specific framework conditions relevant for the development of world-class clusters. http://www.clusterobservatory.eu/index.html#!view=documents;mode=one;sort=name;uid=1b5a1d43-1eb4-42c0-b1fa-ca136d5551e2;id=. Accessed on 16 Feb 2016

European Commission [EC] (2010a) Green paper: unlocking the potential of cultural and creative industries. European Commission, Brussels

European Commission [EC] (2010b) European Competitiveness Report 2010. European Union, Luxembourg

European Commission [EC] (2011) Creative Europe (2014–2020) – a new framework for the cultural and creative sectors. Communication from the Commission to the European Parliament, the Council, the European Social and Economic Committee and the Committee of Regions, Brussels, Belgium, 23 November 2011

European Commission [EC] (2012) Communication from the commission to the European parliament, the council, the European economic and social committee and the committee of the region. Promoting cultural and creative sectors for growth and jobs in the EU. COM (2012) 537 Final, Brussels, Belgium, 26 Sept 2012

European Commission [EC] (2014) For a European Industrial Renaissance. Communication from the Commission to the European Parliament, the Council, the European Economic and Social Committee and the Committee of the Region, Brussels, Belgium, 22 Jan 2014

European Commission [EC] (2016) Boosting the competitiveness of the cultural and creative industries for growth and jobs. European Commission, Brussels

European Creative Industries Alliance [ECIA] (2014) Create, innovate, grow. A new policy agenda to maximise the innovative contribution of Europe's creative industries. Recommendations from the policy learning platform of ECIA. http://www.eciaplatform.eu/wp-content/uploads/2014/11/ECIA_report_Create-Innovate-Grow-1.pdf. Accessed 1 Sept 2016

European Parliament [EP] (1999) Cultural industries and employment in the countries of the European Union. Summary. Education and culture series EDUC 104A http://www.europarl.europa.eu/workingpapers/educ/104aensum_en.htm. Accessed on 16 Feb 2016

European Parliament [EP] (2013) European cultural and creative sectors as sources of economic growth and jobs. European Parliament resolution of 12 September 2013 on promoting the European cultural and creative sectors as sources of economic growth and jobs (2012/2302 (INI)). http://www.europarl.europa.eu/sides/getDoc.do?pubRef=-//EP//NONSGML+TA+P7-TA-2013-0368+0+DOC+PDF+V0//EN. Accessed on 15 Feb 2015

European Union [EU] (2012) Policy handbook on how to strategically use the EU support programmes, including Structural Funds, to foster the potential of culture for local, regional and national development and the spill-over effects on the wider economy? Working Group of EU Member States Experts (Open method of coordination) on cultural and creative industries.

http://ec.europa.eu/culture/library/publications/cci-policy-handbook_en.pdf. Accessed on 15 Feb 2016

European Union [EU] (2013) Creative Europe: the European Union programme for the cultural and creative sectors. Available online: http://ec.europa.eu/programmes/creative-europe/documents/creative-europe-flyer_web_en.pdf. Accessed on 2 May 2014

EUROSTAT (2007) Cultural statistics. European Union, Belgium

EUROSTAT (2011) Cultural statistics. European Union, Belgium

EUROSTAT (2016) Cultural statistics. European Union, Belgium

Evans G (2009) Creative cities, ceative spaces and urban policy. Urban Stud 46:1003–1040

Fleming T (2015) Cultural and creative spillovers in Europe. Report on a preliminary evidence review. Fleming Tom Creative Consultancy, http://www.artscouncil.org.uk/advice-and-guidance/browse-advice-and-guidance/cultural-and-creative-spillovers-europe. Accessed on 16 Feb 2016

Flew T (2013) Global creative industries. Wiley, New York

Florida R (2002) The rise of the creative class: and how it's transforming work, leisure, community and everyday life. Basic Books, New York

Florida R, Tinagli I (2004) Europe in the creative age. Carnegie Mellon Software Industry Centre and Demos. https://www.creativeclass.com/rfcgdb/articles/Europe_in_the_Creative_Age_2004.pdf. Accessed 1 Sept 2016

Haase A, Steinführer A, Kabisch S, Grossmann K, Hall R (eds) (2011) Residential change and demographic challenge. The inner city of East Central Europe in the 21st century. Ashgate, Burlington

Hall P (2000) Creative cities and economic development. Urban Stud 37(4):639–649

Hesmondhalgh D (2007) The cultural industries, 2nd edn. Sage, London

Jakob D, Van Heur B (2015) Editorial: taking matters into third hands: intermediaries and the organisation of the creative economy. Reg Stud 49(3):357–361

Kasprzak R (2015) Creative industries in the Polish economy. Quat Geochronol 34(2):35–43

KEA (2006) The economy of culture in Europe. European Commission, Brussels

KEA (2009) The impact of culture on creativity. A study prepared for the European Commission. http://www.keanet.eu/docs/impactculturecreativityfull.pdf. Accessed on 8 Jan 2016

Ketels C, Protsiv S (2014) European Cluster Panorama 2014. Prepared for the European cluster observatory. http://ec.europa.eu/growth/smes/cluster/observatory/cluster-mapping-services/cluster-mapping/cluster-panorama/index_en.htm. Accessed on 8 Jan 2016

Klasik A (2009) Kreatywne miasto – kreatywna aglomeracja. Wydawnictwo Naukowe Akademii Ekonomicznej, Katowice

Kloudouvá (2009) Kreativni ekonomika a jeji měřeni. Ekonomický časopis 57(3):247–262

Kovács Z, Egedy T, Foldi Z, Keresztély K, Szabó B (2007a) Budapest: from state socialism to global capitalism. ACRE report 2.4. AMIDSt, Amsterdam

Musterd S, Kovacs Z (eds) (2013) Place making and policies for competitive cities. Wiley, Chichester

Musterd S, Murie A (eds) (2010) Making competitive cities. Wiley, Chichester

Nurković R (2015) Contemporary development of creative industries in Bosnia and Herzegovina. Quat Geochronol 34(2):45–51

O'Connor J (2010) The cultural and creative industries: a literature review, 2nd ed. Creativity, Culture and Education, Newcastle.http://www.creativitycultureeducation.org/wp-content/uploads/CCE-lit-review-creative-cultural-industries-257.pdf. Accessed 24 Aug 2012

Open Method Coordination Expert Working Group on Cultural and Creative Industries [OMC-EWG] (2010) Final report of the OMC-EWG on maximising the potential of the cultural and creative industries in particular that of SMEs. http://ec.europa.eu/culture/policy/strategic-framework/documents/cci-report_en.pdf. Accessed 6 Sept 2016

Open Method of Coordination Working Group of EU Member States' Experts [OMC-EWG] on Cultural and Creative Sectors (2014) Good practice report on the cultural and creative sectors' export and internationalisation support strategies. European agenda for culture. Workplan for

culture 2011–2014 http://ec.europa.eu/culture/library/reports/eac-omc-report-ccs-strategies_en.pdf. Accessed 6 Sept 2016
Power D (2011) Priority sector report: creative and cultural industries. European Cluster Observatory, Europa Innova Paper N. 16, European Commission, Brussels
Rumpel P, Slach O, Koutsky J (2010) Creative industries in spatial perspective in the old industrial Moravian Silesian region. E+M Ekonomie Manag 4:30–46
Ságvári B, Lengyel B (2008) Creative Atlas. Spatial and temporal change of the Hungarian creative labour. Demos Hungary Foundation, Budapest
Sailer-Fliege U (1999) Characteristics of post-socialist urban transformation in East Central Europe. GeoJournal 49(1):7–16
Stanilov K (ed) (2007) The post-socialist city: urban form and space transformations in Central and Eastern Europe after Socialism. Springer, Heidelberg
Stenning A. (2004) Urban change and the localities. In: Bradshaw M, Stenning A (eds) East Central Europe and the former Soviet Union. The post socialist states. DARG Regional Development Series No. 5. Pearson Education, Harlow
Stryjakiewicz T (2010) Location factors of the creative and knowledge-intensive industries in European metropolitan regions. Geografický Časopis 62(1):3–19
Stryjakiewicz T, Burdack J, Egedy T (2010a) Institutional change and new development paths: Budapest, Leipzig, Poznan, Riga and Sofia. In: Musterd S, Murie A (eds) Making competitive cities. Wiley/Blackwell, Oxford
Stryjakiewicz T, Kaczmarek T, Męczyński M, Parysek JJ and Stachowiak K (2010b) Policies and strategies in Poznan. How to enhance the City's competitiveness. ACRE report 10.8. AMIDSt, Amsterdam
Stryjakiewicz T, Stachowiak K, Męczyński M (eds) (2012) Creative class, creative industries and creative cities. Quat Geochronol 31(4)
Stryjakiewicz T, Męczyński M, Stachowiak K (2014) Role of creative industries in the post-socialist urban transformation. Quat Geochronol 33(2):19–37
Stryjakiewicz T, Churski P, Stachowiak K (eds) (2015) Cultural and creative industries in the development policies of cities and regions. Quat Geochronol 34(2)
Švob-Đokiæ N (ed.) (2007) Cultural transitions in southeastern Europe. The creative city: crossing visions and new realities in the region. Culturelink Joint Publication Series No 11. Institute for International Relations, Zagreb
Sztompka P (1995) Cultural and civilisational change: the core of post-communist transition. In: Grancelli B (ed) Social change and modernisation: lessons from Eastern Europe. Walter de Gruyter, Berlin
TERA (2010) Building the digital economy. The importance of saving jobs in the EU creative industries. https://oami.europa.eu/ohimportal/documents/11370/71142/Building+a+Digital+Economy,+the+importance+of+saving+jobs+in+the+EUs+creative+industries. Accessed 21 Dec 2014
TERA (2014) The economic contribution of the creative industries to EU GDP and employment. Evolution 2008–2011. http://www.teraconsultants.fr/en/issues/The-Economic-Contribution-of-the-Creative-Industries-to-EU-in-GDP-and-Employment. Accessed 21 Dec 2014
Throsby D (2008) Modelling the cultural industries. Int J Cult Policy 14(3):217–232
Tomczak P, Stachowiak K (2015) Location patterns and location factors in the cultural and creative industries. Quat Geochronol 34(2):7–27
Trippl M, Todtling F, Schuldner R (2013) Creative and cultural industries in Austria. In: Cooke P, Lazzaretti L (eds) Creative cities, cultural clusters and local economic development. Edward Elgar, Cheltenham
Tsenkova S, Nedović-Budić Z (eds) (2006) The urban mosaic of post-socialist Europe, Space, institutions and policy. Physica Verlag, Heidelberg
United Nations [UN] (2008) Creative economy report 2008. The challenges of assessing the creative economy: towards informed policy making. UN, Geneva/New York.

United Nations [UN] (2010) Creative economy report 2010. Creative economy: a feasible development option. UN, Geneva/New York.
United Nations [UN] (2013) Creative economy report 2013. Special edition. Widening local development pathways. UN, Geneva/New York
Węcławowicz G, Potocka M, Baucz A (2010) Rozwój miast w Polsce. Ministerstwo Rozwoju Regionalnego, Warszawa

Part I
Interplay Between Sectoral and Spatial Dynamics of the Creative Industries

Chapter 2
Specialisation as a Driver of the Development Dynamics of Creative Cities and City Regions

Krzysztof Stachowiak and Tadeusz Stryjakiewicz

2.1 Introduction

In this chapter we try to explore how cultural and creative industries (CCIs) can contribute to the economic development of cities and city regions through specialisation. In other words, we want to answer the question: to what extent and in what conditions specialisation can become a real driver of urban development dynamics? The role of specialisation is confronted here with that of diversification, which – in turn – contributes to a long-lasting discussion on the mutual interdependence (or opposition) between Marshall-Arrow-Romer and Jacobs externalities.

The chapter offers both theoretical and conceptual considerations based on an overview of the literature and empirical evidence taking the form of case studies. The chapter consists of two main parts. In the first one (Sect. 2.2), the key issues of the current debate on specialisation of creative cities and regions are presented. In the second part (Sect. 2.3), we distinguish three types of creative cities and regions' specialisation and illustrate them with case studies from different urban areas across Europe. Finally we conclude that specialisation can be a driver of the creative development dynamics not only in large metropolitan areas but also in small and medium-sized cities.

K. Stachowiak (✉) • T. Stryjakiewicz
Institute of Socio-Economic Geography and Spatial Management, Adam Mickiewicz University, Poznań, Poland
e-mail: krst@amu.edu.pl; tadek@amu.edu.pl

© Springer International Publishing AG 2017

C. Chapain, T. Stryjakiewicz (eds.), *Creative Industries in Europe*,
DOI 10.1007/978-3-319-56497-5_2

2.2 Key Issues of the Debate on Specialisation of Creative Cities and City Regions

The debate on specialisation of creative cities and city regions embraces many perspectives and key issues. In this section, we discuss the most important ones from the point of view of the topic and conceptual background of this chapter. The first issue relates to a controversy about the advantages of economic specialisation versus diversification in urban development. The second issue concerns the conditions which favour the potential for urban creative specialisation. This potential consists of both current existing factors supporting the growth of cultural and creative industries (described by the concept of creative field or creative milieu), and factors inherited from the past (described by the concept of path dependence). The third element of our discussion is an attempt at identifying the main factors contributing to specialisation, to which we have included industrial profiles and location patterns. At the end of this section we offer to look beyond the diversity-specialisation divide and introduce the concepts of diversified specialisation and a diversified-specialised creative city. These concepts are the base of the typology of specialisation of creative cities presented in Sect. 2.3.

2.2.1 Specialisation vs. Diversification of Creative Cities and Regions

For decades, geographers, economists and urban theorists have debated the role of economic specialisation and diversification in economic development. On the one hand, starting with the theory of comparative advantage in the early nineteenth century, the benefits of specialising in "what one does best" have been emphasised. On the other, it has been argued that the diversification of production can make a city less prone to negative economic shocks. Yet, there is still no straightforward answer as to whether urban specialisation and Marshall-Arrow-Romer externalities (Marshall 1920; Arrow 1962; Romer 1986), or a diversified economic structure and Jacobs externalities (Jacobs 1969), foster economic well-being and growth.

In the recent years, the world's richest cities and city regions have been those whose economies feature a concentration of employment in information technology and finance. In the developed countries, big manufacturing regions are in decline in terms of their income rank and often in their population, while in the developing world, hubs of export-oriented, labour-intensive manufacturing, such as Guangzhou in China, are said to have the secret of growth. Specialisation is the principal way, then, in terms of which urban economies are viewed, labelled and classified by practitioners and policy-makers, and it defines the public image of specific cities. This is one of the reasons why Storper (2013: 32) calls specialisation "the motor of urban economies", and Sassen (2009) observes that specialised cities make a difference.

Many scholars point out that the diversity of a city's economy is crucial to the growth of cultural and creative industries (CCIs) (Landry 2000; Florida 2002b, 2005; Hall 2004; Krätke 2011). However, some argue that it is not always enough. Scott (2006a), for instance, adds labour specialisation and cultural production. Labour specialisation, according to this author, requires an agglomeration (or a spatial concentration) to achieve the flexible specialisation in order to compete in the new global economy (Scott 2006a: 3). In addition, he argues that there should be a balance between these production systems and urban cultural assets (2006a: 10). The combination of the two tends to give larger cities an advantage, but there are notable exceptions where specialised industries exist. Scott (2006a) notes that there are many small and specialised creative cities and agglomerations all over the world. The continued survival of such places "can in part be accounted for by the powerful economic advantages conferred on them by the articulation of specialised agglomeration processes with insistent product differentiation. These advantages allow them in many cases to flourish notwithstanding the competitive pressures that flow from larger and vastly better endowed centres of production" (Scott 2006a: 9).

The relationship between the level of specialisation (expressed by the location quotient LQ) and the characteristics of city regions indicates that specialisation is more likely to occur in bigger, more densely populated, more urbanised and more prosperous city regions (Table 2.1). These regions tend to specialise in such industries as advertising, radio and TV, or software. In more generic activities, such as retail and the distribution of creative goods, or museums and the preservation of historical sites and buildings, it is hard for any cities or regions to specialise. Those activities might occur even in smaller and less prosperous areas. As Table 2.1 indicates, museums and heritage activities are fairly concentrated, therefore they might serve as a potential specialisation niche for more peripheral regions.

At the micro level, a so-called specialisation effect might contribute to innovation. As is widely known, most creative enterprises are very small, which may result in some constraints. Limited resources make it difficult to allocate sufficient effort to the various activities needed to run a business. Specialising in a certain creative activity (e.g. designing computer games) and purchasing complementary creative inputs needed to sell a final product (e.g. software programming) may allay this capacity constraint. Müller et al. (2009), who conducted a survey of more than 2000 creative industry enterprises, point out that creative enterprises using inputs from other creative firms turn out to be significantly more capable of introducing innovations and conduct in-house R&D than those that refrain from reaching for creative products and services from outside. This is what they call a specialisation effect. Purchasing creativity from other enterprises allows a firm to concentrate on its own strengths. Teamwork, i.e. the establishment of a network of creative enterprises to deliver products and services jointly, affects innovation activities only in the case of stable networks.

Table 2.1 Correlation between specialisation and regional features in selected cultural and creative industries

	Coefficient of correlation between specialisation (LQ) and			
Cultural and creative industries	population	population density (people per sq km)	degree of urbanisation (%)	per capita GDP (EUR PPP)
Advertising	0.23**	0.37**	0.45**	0.56**
Artistic creation and literary creation	0.16**	0.36**	0.38**	0.45**
Museums and preservation of historical sites and buildings	−0.01	0.16**	0.14*	−0.14*
Printing and publishing	0.07	0.34**	0.44**	0.49**
Radio and television	0.26**	0.57**	0.48**	0.51**
Retail and distribution of creative goods	0.01	0.18**	0.26**	0.74
Software	0.16**	0.23**	0.53**	0.49**

Source: Own calculations based on European Cluster Observatory data
Note: Data for 255 NUTS2 European regions
**$p < 0.01$
*$p < 0.05$

2.2.2 *Potential for Urban Creative Specialisation: The Creative Field and Path Dependence*

The relation between creative activities and the cities in which they are located is that of mutual influence and dependence. This kind of relation, which is often complex and multidimensional, lies at the very heart of our understanding of creativity-driven, urban economic development and creative activities themselves (cf. Landry 2000; Florida 2005; Pratt 2008; Musterd and Murie 2010; Krätke 2011; Scott 2014). Cities are regarded as places where creativity is generated, applied and, at the same time, influenced by other creative activities that occur in the same space. A set of resources, often of an immaterial nature, is available in specific cities, or within sectoral and personal networks that entwine them. Those resources are used by both creative individuals and organisations. All this generates a complex dynamics of intertwined factors, which can be called a **potential for urban creative specialisation**. It consists of two elements: (1) current conditions for the growth of cultural and creative industries, forming a so-called creative field or creative milieu, and (2) past conditions which determine both current and future capabilities for creative and cultural industries (path dependence).

According to Scott (1999, 2006b, 2010), the city is a powerful fountainhead of creativity, and it can be understood in terms of a series of localised field effects. **A creative field** consists of four major components, namely, (a) intra-urban webs of specialised and complementary producers; (b) the local labour market and the social networks that bind workers together in urban space; (c) the wider urban environment, including various sites of memory, leisure and social reproduction;

and (d) institutions of governance and collective action. The concept of a creative field stresses the role of economic activities based on creativity in an urban environment and argues that it depends on various factors that can influence each other. A creative field "is represented by sets of industrial activities and related social phenomena forming geographically-differentiated webs of interaction giving rise to diverse entrepreneurial and innovative outcomes. An intrinsic element of this definition is that both the field on the one side and its effects on entrepreneurship and innovation on the other are reflexively intertwined with one another" (Scott 2006b: 3). This suggests that there is a shared relational context between creative actors in a given place, as well as between them and the local socio-institutional context, so these relations cannot be reproduced elsewhere. Creative actors bring in their human capital and creativity, but they also learn norms, strategies, and organisational forms from the context, or milieu, in which they operate that can also influence production. As Scott (2010: 119) writes, "the individuals who compose each community typically internalize elements of their daily environment and reflect these back in more or less socially conditioned creative efforts". This kind of knowledge is mostly related to urban regions, and thus is influenced by local entrepreneurial characteristics and actions, as well as by professional networks at the local level.

The conceptual basis of the creative field is not entirely new, and some elements of it can be found in the literature on such topics as the innovative milieu, the learning region, the regional innovation system and the like. The predominant concern of those concepts has been the technology-intensive industry, and the concept of the creative field shifts the focus to symbol-intensive production in the cultural economy. As Scott (2010: 122) notes, an important consequence of this shift of emphasis is the need to pay very special attention to the cultural resonances of place. In this case a creative field is an extension of what is known as a **creative milieu**. According to Hall (1998), it was the French philosopher Hippolyte Taine who introduced the concept of the creative milieu in 1865 (Taine 1873). In the geographical literature it was independently introduced by Törnqvist (1983) and examined by Aydalot (1986) or Camagni (1995) as an innovative milieu. A creative milieu is a set of resources existing in a particular place that form a favourable environment for creative and innovative activities. Those resources are: a large body of information readily transferable within such an area, a body of knowledge based not only on the accumulation of data, but also on the accumulation of knowledge over time (as in university centers), and expertise in specified types of activity. The combination of those three types of resources is a condition for the emergence of a fourth, namely creativity, which means an ability to create new forms and values, whether material (e.g. products) or immaterial (e.g. signs, ideas, brands) (Stryjakiewicz 2010). Further work on the creative milieu concept has added more resources to those mentioned above (Meusburger et al. 2009; Cusinato and Philippopoulos-Mihalopoulos 2016). In principle, the creative milieu concept implies that in order to generate creative and innovative outputs a specific set of intertwining factors have to occur in a particular space. As Meusburger (2009: 111) points out, "single variables or models ignoring spatial contexts apparently have less power to explain creativity than some spatial categories do. The various personal, organizational, material, cultural, and political factors affecting creativity interact, merge, and modify each other in specific places or areas and lead to spatially rooted macrophenomena called milieu, environment, action setting, context, or knowledgescapes". According to Florida (2002a), a creative milieu is a habitat in which interconnected technological, entrepreneurial and artistic types of creativity stimulate one another. Such a specific social and cultural environment attracts new creative people and supports a dynamic transfer of knowledge and ideas. There are two chief aspects of a creative milieu: (1) the concentration of creative capital and (2) an accumulation of diversity. Here the term "concentration" does not mean only a large number of people, but primarily the frequency and quality of interactions taking place among them. Diversity, in turn, refers not only to a cultural variety, but also to people's activities and city functions (Meusburger 2009; Rumpel et al. 2010).

The creative field or creative milieu and economic development related to it reflect the **path dependence** of cities. The economic profile and innovation capacity of a city are determined by the main industries located there (Hall 1998; Scott 2008; Storper 2013). In this sense, the development of a sector, the related specialisations, and the labour market influences local creative energies and possible innovation trajectories. Consequently, the need for the recognition of creative results in one place can drive creative actors present in a given city sector, because those kinds of creative inputs are sought after and acknowledged. Staber (2008) notes that cultural regions often specialise the way firms do. In order to build a competitive advantage, the ideas occurring in a region should be consistent with core ideas defining the region. "The reason for this is that an environment filled with semantic associations, symbolic meanings and cultural representations that reflect the core set of ideas can trigger people to perceive a new idea as being consistent with that set than when such associations are absent" (2008: 573). Parts of the creative field are linked to the urban production system and also generate dynamism on the local labour market (Storper and Scott 2009). Hence, in sum, the creative field of a city is unique and, even if it does not coincide spatially with the city region, its profile is defined by the result of an interaction between economic actors and physical and socio-institutional elements.

Hall (2004) points out that an important element to understand when looking at a creative city is how its history influences its current economic situation and dynamics. The research on local and regional trajectories and path dependence is becoming a broad conceptual framework in social sciences, but its definition and application often remains rather vague (Hausner 1995; Deeg 2001; Tissot and Merger 2001; Fuchs 2005). In the broadest sense, the concept refers to the notion that "history matters". While the importance of history can hardly be overstated when explaining the present and exploring the future, path dependence should have a more concrete meaning if it is to be useful in the analysis of urban and regional dynamics. Pierson (2000) and Mahoney (2000) offer narrower and more precise definitions that can form a starting point for research. Pierson (2000: 252) proposes the idea of increasing returns: "the probability of further steps along the same path increases with each move down that path". Mahoney (2000: 507) states that "path dependence characterizes specifically those historical sequences in which contingent events set into motion institutional patterns or event chains that have deterministic properties". Economists and economic geographers have studied path dependence in the creation and reproduction of economic branches and networks of interrelations at certain locations, often in the shape of specific clusters (Boschma and Kloosterman 2005). According to Bontje et al. (2011), two applications of the path dependence concept are significant for research on the creative economy. The first relates to the role of path dependence in the creation and reproduction of businesses in particular industries and networks of interrelationships at certain locations, which contribute to the formation of clusters. The other path-dependence approach refers to the evolutionary paths of cities and regions and the consequences of those paths for their current and future development. Many cities and regions, especially with a manufacturing legacy, have recently been, and still are, searching for a new economic profile to stay or become domestically and internationally competitive. Over time it leads to the development of production networks and their stronger territorial embeddedness.

Specialisation is more likely to occur in cities or regions with a strong presence of a creative field, or where a certain creative or cultural industry was an important element of their development path. Those two conditions – a strong creative field along with the tradition of a particular creative activity (such as violin making in Cremona, Italy) – may facilitate the current specialisation of a city or region. It is important to note that the co-existence of a creative field and tradition is important. A rich past together with a present creative talent and potential add to a strong creative city or region. In the empirical part of this article (Sect. 2.3) we will present some examples in detail.

2.2.3 *Industrial Profile and Location Patterns of Cultural and Creative Industries as Factors Contributing to Specialisation*

The earlier development path of a city or region leads to the formation of its present identity and industrial profile, as explained in the previous section. The creative field in this context can be considered an internal factor contributing to the specialisation in cultural and creative activities. In addition, there is also an external factor which does not depend on place specificity, but rather on the ways an industry operates and how those operations are organised across space. Therefore location patterns of cultural and creative industries can be regarded as an important force shaping the creative specialisation of a city or region.

Although cultural and creative industries are often analysed as a single group, they tend to differ from one another. The nature of a creative activity within particular creative and cultural industries varies and may be located in space in different ways (Tomczak and Stachowiak 2015). Using the example of the Italian creative sector, Bertacchini and Borrione (2013) carried out a research chiefly intended to give an insight into how the features of various creative industries affected their location patterns. Although it is generally believed that while service-oriented creative industries tend to locate in metropolitan areas owing to their customer orientation, those connected with design exhibit a more diversified location pattern. These differences, according to those authors, follow from the double nature of design. On the one hand, those branches have a strong productive character based on traditional craftsmanship or industrial techniques. On the other, the designing process and innovativeness are significant factors in creating their non-material values and giving them a load of the local culture.

As to tendencies in the distribution of enterprises of cultural and creative industries, they are usually said to concentrate in urban areas and form specialised clusters (Turok 2003; Wu 2005; Cooke and Lazzeretti 2008; Lazzeretti et al. 2008; Evans 2009; Chapain et al. 2010). Table 2.2 shows the concentration of cultural and creative industries in Europe. A generally high propensity for spatial concentration can be explained by urbanisation effects that follow from interactions among various sectors and activities, and that provide a counterbalance to the higher transport and operational costs in cities than if an enterprise was located in a peripheral area. Among significant urbanisation effects Rumpel et al. (2010) list the following: proximity of customers, proximity of suppliers, access to information, and a creative milieu.

Creative firms mostly produce one-off commodities or services, the final shape of which emerges in the course of close cooperation with customers. This interaction between firms and customers requires a relatively small geographical distance allowing frequent face-to-face contacts, in particular in the initial and final stages of the implementation of a project. The geographical proximity of a wider circle of recipients also reduces the high risk of customer loss characteristic of the creative sector. In turn, customers choose a supplier of creative goods and services from a larger pool than would be possible in peripheral areas. Spatial proximity can also have negative effects, as illustrated by the example of Leipzig, discussed in detail in Sect. 2.3.2 of this chapter. An excessive orientation towards the local market can lead to stagnation caused by too strong links between suppliers and customers and a "lock-in" effect.

When speaking of the proximity of suppliers, it is worth emphasising that creative commodities and services are usually made by designing teams set up for a specified time. Some of them require specialised knowledge, often surpassing the competence of a single firm. The spatial proximity of other enterprises, also non-CCI ones, allows on the one hand a greater flexibility of response in the case of a dissolution of existing business relations, and on the other, prevents the so-called paradox of embeddedness (Uzzi 1997), which can be explained by a simple mechanism. While ever new possibilities of profitable transactions with alternative partners keep turning up, enterprises will not want to be bound by long-term contracts with just one. However, when this type of opportunistic behaviour leads to a situation in which the uncertainty of exchange escalates and the risk in the

Table 2.2 Concentration of cultural and creative industries in Europe in 2010

Sub-sectors of cultural and creative industries	Gini index
Publishing of computer games	0.91
Manufacture of magnetic and optical media	0.83
Reproduction of recorded media	0.82
Motion picture, video and television programme distribution activities	0.78
Motion picture, video and television programme post-production activities	0.75
Operation of historical sites and buildings and similar visitor attractions	0.73
Web portals	0.71
News agency activities	0.70
Television programming and broadcasting activities	0.69
Cultural education	0.69
Library and archives activities	0.66
Sound recording and music publishing activities	0.66
Other software publishing	0.66
Manufacture of musical instruments	0.66
Radio broadcasting	0.64
Printing of newspapers	0.63
Museums activities	0.61
Operation of arts facilities	0.61
Motion picture, video and television programme production activities	0.61
Specialised design activities	0.57
Other publishing activities	0.55
Artistic creation	0.55
Media representation	0.54
Book publishing	0.53
Binding and related services	0.51
Publishing of journals and periodicals	0.51
Support activities to performing arts	0.49
Translation and interpretation activities	0.49
Retail sale of music and video recordings in specialised stores	0.47
Performing arts	0.47
Renting of video tapes and discs	0.45
Computer programming activities	0.43
Publishing of newspapers	0.40
Retail sale of newspapers and stationery in specialised stores	0.38
Architectural activities	0.37
Advertising agencies	0.36
Pre-press and pre-media services	0.36
Photographic activities	0.34
Retail sale of books in specialised stores	0.31
Motion picture projection activities	0.30
Other printing	0.23

Source: Power (2011: 28), modified

Note: Due to data availability 129 regions in 17 countries are included: Austria, Belgium, Cyprus, Denmark, Estonia, Finland, France, Germany, Ireland, Italy, Luxembourg, Latvia, Lithuania, Portugal, Slovenia, Sweden and Switzerland.

transactions contracted is considerable, firms tend to think less of changing to more advantageous partners, which leads to their cooperation with a single one becoming too close. Rather than considering new transactions with strangers, more profitable but open to some risk, entrepreneurs prefer regular contacts with well-known actors (Lazzarini et al. 2008).

The success of firms largely depends on constant access to information and knowledge. In the literature on cultural and creative industries it is generally accepted that individual units gain information via the so-called "local buzz" (Bathelt et al. 2004). It rests on both a system of regular long-distance communication and on face-to-face meetings of persons from various firms within the same sector, area or region. Grabher (2002: 254) suggests that a more adequate term to be used in this context is "noise", since it best conveys "the concoction of rumours, impressions, recommendations, trade folklore, and strategic misinformation" on the basis of which enterprises not only make decisions, but also monitor competitors.

In addition to the urbanisation economies explained above, the overall structure of an industry and the way it operates can be a decisive factor in a city's or region's creative specialisation. Cultural and creative industries are currently under a global shift (Flew 2013). Due to rapid digitalisation, both production and distribution of creative goods have been altered and become globalised. This has changed the dynamics of many industries, influencing also the places where those activities are located, and contributing to their specialisation. A good example of a creative industry undergoing a global shift is the film industry. It is organised around various stages in the production chain: a preparatory stage (pre-production), shooting (production) and post-production. Some researchers also add the stages of the distribution of a film and its projection. While traditionally almost the entire chain of film production used to be concentrated in a single place (e.g. in Hollywood), today a growing number of stages are introduced in a variety of locations (Walls and McKenzie 2012; Mirrlees 2013). In fact, film production has recently emerged as a global production network. The term "global" does not necessarily imply that such a network actually spans the entire world; rather, it suggests that it is geographically extensive and functionally integrated across national boundaries. As such, globalisation of the film industry involves the expansion of production away from its traditional centres, whether to other countries or to other locations within the same country. This has been reinforced by recent trends in the film industry, like cross-border film production or the rise of production networks through international co-production initiatives which affect established production locations (Dahlström and Hermelin 2007). Some groups, notably from the US film industry, refer to this development as a "runaway production" (Johnson-Yale 2008). While it is now generally agreed that the film industry is turning into a global network, there is a dispute among scholars on whether the nodes of this network tend to become less important than the linkages (Wasko and Erickson 2008). Nevertheless, the majority of film production industries are concentrated in a relatively small number of specialised places called film or media clusters (Karlsson and Picard 2011b). This is visible in Table 2.2, where motion picture and media sub-sectors tend to be among the most concentrated cultural and creative industries in Europe. A new landscape of the global film production includes: (1) the global spread of film production infrastructure, such as film studios and film clusters, (2) the emergence of cost-cutting incentives, or policies intended to attract international film production, (3) the rise of international collaboration in this field, (4) the increased global interconnectedness of film production companies, and (5) the emergence of satellite production centres. The case of the film industry shows that the location pattern of creative enterprises depends in particular on the branch of industry they belong to, and on whether their activity is divided into stages in the production chain.

2.2.4 *Beyond the Diversity-Specialisation Divide: Diversified Specialisation*

Large cities offer a combination of urbanisation economies as well as so-called Jacobs externalities that stem from the coexistence of diverse industries (Duranton and Puga 2000). If we take into account the innovation regime of creative industries which is based more on a combination of the existing knowledge rather than on new knowledge in terms of scientific discoveries and technology transfers (Asheim et al. 2007), diversity can be considered a key factor for creative industries. Nevertheless, the important fact is that large cities offer not only urbanisation advantages but also partial field specialisations that are significant for higher efficiency in incremental innovations (Lorenzen and Frederiksen 2008). Therefore the concept of "diversified specialisation" has been used to bridge the specialisation-diversity duality (Farhauer and Kröll 2012).

Farhauer and Kröll (2012) argue that one of the major weaknesses inherent in the dichotomy of specialisation and diversification is that cities are only classified into specialised and diversified ones, while other possible sector structures are excluded. Therefore, they have introduced the term "diversified specialisation" which refers to specialisation in more than one, i.e. a few, sectors. "Many cities exhibit multiple specialisations, but – apart from specialisation in a few sectors – they show a diversified structure at the same time" (Farhauer and Kröll 2012: 67).

In **diversified cities**, companies can benefit from Jacobs externalities. These externalities come from a diversified surrounding as new ideas may be generated via knowledge spillovers between different sectors. As soon as the sectors have found their optimal production process, they benefit from a diverse surrounding only to a lesser extent. Therefore, it may pay off to resettle from a diversified to a **specialised city**. First, specialised cities tend to be smaller than diversified ones (Duranton and Puga 2000), which is why agglomeration costs are usually lower in specialised cities, and firms can save costs in moving. Secondly, Marshall externalities gain in importance for firms that have already found their best production process. With Marshall externalities, innovations lowering process costs become more likely, since firms in specialised cities share almost the same knowledge base.

The distinct advantage of **diversified-specialised cities** is that firms in those cities profit from both Marshall and Jacobs externalities. Thus, productivity levels and productivity growth rates can be higher than in fully diversified or fully specialised cities. Companies benefit from the proximity to other firms. Moreover, firms profit from tailored infrastructure and a qualified specialised labour pool. Besides, a (limited) variety of sectors generates cross-fertilisation and higher productivity growth rates than in specialised cities. In diversified-specialised cities, spillovers can said to be "co-ordinated", as they occur between related sectors (Farhauer and Kröll 2012).

One of the reasons for diversified specialisation likely to occur in creative cities and regions is the **co-location of cultural and creative industries**. We can distinguish two types of co-location patterns: (a) a shared location of various creative activities (e.g. the co-location of film, TV and advertising) and (b) the co-location of cultural and creative industries and non-cultural and creative industries activities (e.g. design and manufacturing). And it is the case both at the city and regional levels. Currid and Williams (2010) found that several cultural subsectors showed strong co-location patterns. Using very detailed data for Los Angeles and New York, the authors calculated correlation coefficients across districts at 0.75 and higher for (1) performing arts and music, (2) music and film, (3) art and design and (4) art and film. The co-location patterns are explained by cultural infrastructure. Also Power (2011) shows, with reference to European regions, that advertising as well as radio and television activities are likely to co-locate in the same regions (Table 2.3). Design shows a tendency to co-locate with advertising, and to a lesser extent with games publishing. Artistic and literary creation employees tend to be over-represented in those regions where radio and television activities are over-represented too. Power suggests that such relationships can indicate that firms in certain sectors see positive synergies from locating their staff near one another, perhaps because certain sectors work intensively together, and that co-location adds to business competitiveness and employee productivity. Alternatively, it can be because individuals with certain knowledge and skills are drawn to particular regions, and because firms in need of such skills are drawn to those regions. Those patterns are especially visible in media clusters (for detailed examples see Karlsson and Picard 2011b).

The second type of co-location which is an important factor contributing to the diversified specialisation of a city or region is the location of certain cultural, creative and other industries. It is important to note that relationships between creative and non-creative industries and their spatial consequences are often complex. Design might serve as an example. The design industry, once playing a supporting role for other activities, such as manufacturing, is becoming more and more independent, locating in places with strong creative fields and thus contributing to specialisation. Design may follow geographical location patterns similar to or different from those of manufacturing, but interrelations between those two types of activity are crucial for the emergence of creative systems of design-related industries (Scott 1996). Although design often occurs in the same place as manufacturing, there are cases when both industries are disconnected in space.

Table 2.3 Co-location tendencies in five cultural and creative industries in Europe

	Design	Advertising	Artistic	Games	Radio/TV
Design					
Advertising	0.15*				
Artistic	0.05	0.06			
Games	0.12*	0.34**	−0.01		
Radio/TV	0.08	0.29**	0.17**	0.02	

Source: Power (2011: 29)
Note: Figures represent correlation between location quotients in five cultural and creative industries; the table is based on data for 266 European regions
**$p < 0.01$
*$p < 0.05$

Design, whether a responsibility of specialised divisions in a firm or of contracted freelancers, increasingly becomes an independent activity geared towards preparing a project, irrespective of whether or not it is going to be put into production. This activity shows a tendency to locate in the creative milieu of metropolitan areas, unlike manufacturing, which tends to locate in specialised industrial zones (Bertacchini and Borrione 2013).

2.3 Three Types of Specialisation of European City Regions in Cultural and Creative Industries and Their Evidence

Looking at the main factors contributing to the specialisation of creative cities and regions presented in the previous section, we propose the following typology:

1. place-dependent specialisation, which arises from potential urban creative specialisation embedded either in the creative field of a city or deriving from its development path;
2. industry-dependent specialisation, which is related to the city's industrial profile and the location pattern of cultural and creative industries; and
3. diversified specialisation.

Below we present three case studies illustrating basic features of each specialisation type.

2.3.1 Place-Dependent Specialisation: The Case of Cremona, Italy

Located in Lombardy in the northern part of Italy, Cremona is one of the world's greatest violin making clusters. Its history dates back to the sixteenth century when Andrea Amati gave birth to the contemporary violin. The violins made by his protégés, Antonio Stradivari and Giuseppe Guarneri del Gesu, are still played today with honour by top professional performers, adding an essential part to the art of music performance (Wali 2010). From the mid-seventeenth century to the mid-eighteenth century, Cremona maintained its autonomy and flourished as a harbour between Venice and Milan, even though the city had been under the rule of either Milan or Venice in various periods. Stradivari and his colleagues created their soon-to-be-admired instruments during the Golden Age (Wali 2010). However, as soon as Cremona's power faltered, violin makers spread throughout Italy, which led to a difficult period for the city's violin industry. Cremona finally revived as the manufacturing source of violins during the post-Second World War era. This shift was prompted by the foundation of a violin-making school proposed by Mussolini and opened in 1937. This act, however, did not produce immediate results. It was only in the 1970s that the city revived as the place of violin manufacture. Today, approximately one hundred and thirty violin-making studios are officially registered, and the craftsmen are estimated to number seven hundred (Dioli and Rizzi 2015).

Since the sixteenth century the reputation of Cremona is linked to violin making, but there has also been the tradition of producing other stringed instruments there, such as cellos, violas, double basses and organs, especially in nearby Crema. In musical terms, Cremona has also given birth to two great composers, Claudio Monteverdi and Amilcare Ponchielli. All these elements make the city and province of Cremona emanate intellectual stimulation, skills and knowledge that inspire and attract great musicians and violin makers. The city has a particular atmosphere which is an essential element in producing a favourable creative field able to trigger a cross-fertilisation process (Dioli and Rizzi 2015). The presence of violin makers, high-quality schools and the support of local strategies help to continue and foster its rooted tradition: Cremona and its surroundings are still capable of generating remarkable violin makers. The identity of the city is thus associated with excellent handicraft and music in general. All these elements (the making of bow instruments, education and music-related activities) produce multiplier effects characteristic of clusters and stimulate new initiatives, including a project called "Cultural Regions".

"Cultural Regions" is a project of Fondazione Cariplo, a charitable foundation set up in Milan in 1991 to promote the cultural heritage within a broader local development strategy. It is a unique project in Italy which embraces six geographical areas of Lombardy intended to encompass related businesses in the cultural sphere, improve the governance of the cultural heritage at the local level, promote economic and financial sustainability in its management, and improve the skills of local players for better heritage conservation, management and capitalisation (Gugu et al. 2014). In 2006, at the beginning of the process, the Cariplo Foundation co-funded operational feasibility studies throughout the region to help identify distinctive traits of various areas and their potential for making a cultural region. A call for proposals was followed by in-depth feasibility studies. Then a selection was made where the Foundation singled out and approved six cultural regions, giving each of them some funds for cultural actions (Ponzini et al. 2014). Over the years the Cariplo Foundation has provided technical assistance, and monitored and promoted various actions. The total investment in the project has amounted to 65 million euros; grants by the Cariplo Foundation have totalled 20 million euros, the rest coming from local private and public entities operating in each region (Gugu and Dal Molin 2016).

The cultural region in the province of Cremona was established in 2011. By 2013, nine million euros had been invested in it, of which 2.6 million euros donated by the Cariplo Foundation (Ponzini et al. 2014: 81). Music and craftsmanship are central to the local culture and economy of Cremona. The region intends to coordinate the artistic and entrepreneurial initiatives already launched in the music-related creative field, to restore and reuse historic buildings for activities related to music, and to develop strong territorial marketing (Ponzini et al. 2014). The main participants of the project include: the Cremona Chamber of Commerce, eight municipalities, the Stradivari Foundation, Pavia University, artisan training organisations, as well as music and performing arts foundations and associations (Dioli and Rizzi 2015).

2.3.2 *Industry-Dependent Specialisation: The Case of Leipzig, Germany*

From a geographical point of view, one of the most distinctive features of creative industries related to media, such as film, TV, music or video games, is their heavy concentration in a limited number of large cities where big media clusters have developed (Krätke and Taylor 2004). These cities specialising in media are connected by a worldwide network operated by a limited number of large transnational media corporations working at a global scale and controlling a global network of branch offices and subsidiaries. In this way cities become local and national nodes in the global media network. Global media corporations implement strategies focusing on opening up new markets and increasing their market shares, which necessitates their presence in cities specialising in media or in large media centres (Mirrlees 2013). Their presence in those cities allows large media firms to follow the latest trends and developments in cultural and creative industries as well as the latest technological developments affecting the media industry. Around those corporations that function as anchors there develop clusters of small, often very small, specialised media firms engaged as subcontractors to large media firms or in project-oriented productions. These clusters function as creative urban milieux characterised by intensive and flexible forms of inter-firm networking and allowing global media corporations to tap the globally dispersed creative potential of entertainment and audio-visual production (Karlsson and Picard 2011a).

Because of the specificity of the media industry described above, several cities are highly specialised as media centres. There are numerous examples of such cities worldwide, including London, Paris, New York, Los Angeles, Vancouver, or Toronto. They are often called "global media cities" (Krätke and Taylor 2004). However, there are also a few smaller specialised media centres, such as Leipzig. Leipzig is situated in Eastern Germany and has a population of 520 thousand (2015). As a former major commercial, international fair and cultural centre in the heart of Europe, but also as a heavily industrialised city in the former GDR, the city-region has been going through a substantial multi-directional transformation since 1990. The transition to a market economy brought about by German reunification in 1990 quickly led to a widespread collapse of traditional local economic structures. The city had to develop new, future-oriented, knowledge-based branches of the economy and service industries, and had to adapt to the needs of the existing regional workforce. Large investments in transport infrastructure, such as roads, railways and a new runway at the Leipzig-Halle Airport, have led to a further integration of the city region with national and European economic areas (Lange et al. 2007; Stryjakiewicz et al. 2010).

Since the 1990s the media sector has become increasingly important in the local economy. In 1992, the MDR (Mitteldeutscher Rundfunk, the regional broadcasting corporation serving the states of Saxony, Saxony-Anhalt and Thuringia) opened its headquarters in Leipzig. The MDR turned out to be a major hub for the development of a media cluster. Shortly afterwards, a media development agency, Medienstadt Leipzig GmbH, was set up to support the growth of media-related activities (Chapain et al. 2013). The dawn of the Leipzig media industry appears to be a good example of successful industrial clustering. Bathelt (2002) shows how new media firms get started up, how they shape the local labour market, and how this process contributes to the formation of specialised media-related institutions in the region, from which it benefits. The local potential and inter-relationships, such as the development of media-related skills, transaction-specific investments in producer-user relations, and conventions about TV formats, have enabled firms to work out common ground for cooperation and contributed to the growth of the cluster.

Leipzig is also an interesting example of path creation. Bathlet (2002) notes that the growth of the media cluster in Leipzig has not been related to the city's traditional industrial and institutional structures in any way, nor has it resulted from the renewal of a pre-existing industry. The appearance of the cluster has not been embedded into its historical structures and traditional social relations. In this respect, the development of Leipzig contrasts sharply with many other media clusters, such as London, Los Angeles or Toronto.

2.3.3 *Diversified Specialisation: The Case of Prague, Czech Republic*

Diversified specialisation tends to occur in large metropolitan areas. Striking examples include such global media capitals as New York, Los Angeles or London. They are centres for the media, advertising, music, fashion or film (see the London case of the film cluster discussed by Chapain and Stachowiak in this book). What we call diversified specialisation, and what Pratt (2011) labels as "micro-clustering", is not only typical of global cities. There are numerous cities at a national or regional level that tend to specialise in various creative industries. One of them is Prague. As the capital of the Czech Republic, it is the largest city in the country with a population of 1.3 million, and the metropolitan area inhabited by 2.2 million. The city has traditionally been one of the cultural centres of Europe, hosting many cultural events. There are many world-class museums, halls, galleries, cinemas, and music clubs. It hosts many events, such as music and film festivals or fashion shows. According to Němec (2013), the CCIs contributed almost 4.9% to the Czech GVA (cultural industries 1.9% and creative industries 3.0%). The economic significance of the CCIs in Prague was approximately double that of the nationwide average: CCI branches accounted for more than 10% of GVA in Prague, and the capital city itself generated 53.5% of GVA created in the Republic by CCIs. In economic terms, the most significant branches in Prague are information technology and software, the creation of TV and radio programmes, as well as broadcasting and publishing. In 2012 there were a total of 37,617 businesses operating in Prague in the CCI sector, which represented 12.6% of their total number there (Němec 2013: 5). The city's CCIs were of extraordinary significance both in economic terms (measured as a share in GVA) and in terms of employment. The entities under analysis employed almost 125,000 people in Prague, which is an estimated 14% of the city's total workforce (Němec 2013: 5).

The cultural and creative industries in the Czech Republic are significantly concentrated regionally, mainly in large cities with a clear dominance of Prague (Slach et al. 2013). The capital itself accounts for almost 40% of the country's creative sector employment, and its shares in CCI production and CCI value added are even higher (see Table 2.4). The dominance of Prague in the national creative sector is considerably higher than the dominance of capital cities in most other European countries. This dominant position of Prague is also evident when considering the shares of the next two major centres, Brno and Ostrava. These cities account for 9% and 3%, respectively, of all jobs in the creative sector in the country (Slach et al. 2013: 18).

The creative industry which gives Prague international and even global links is film. The Prague film cluster features attractive locations, flexible and highly skilled staff, studios equipped with the latest technologies, a substantial share of foreign productions, a relatively large domestic cinema and DVD market, increased competition from other East European locations, the availability of grants, the pursuit of reforms (e.g. in tax breaks, film law), and the development of a string of related

Table 2.4 Share of Prague in the country's total CCI employment, production and value added in 2009 (%)

NACE codes	18	58	59	60	71	73	74	*CI*	*M*	*T*
Employment	28.2	56.6	57.9	78.3	30.6	51.2	32.3	**38.2**	10.0	23.3
Production	41.3	71.1	81.2	94.8	39.6	74.7	41.7	**54.3**	10.5	28.3
Value added	35.1	67.5	72.0	96.2	38.8	69.8	38.7	**49.3**	12.3	35.1

Source: Slach et al. (2013: 19)
Note: *CI* creative industries, *M* manufacturing, *T* total employment
NACE codes:
18 – Printing and reproduction of recorded media
58 – Publishing activities
59 – Motion picture, video and television programme production, sound recording and music publishing activities
60 – Programming and broadcasting activities
71 – Architectural and engineering activities, technical testing and analysis
73 – Advertising and market research
74 – Other professional, scientist and technical activities

institutions (Borrás and Tsagdis 2008: 194). The flagship organisation of the film cluster is the renowned Barrandov studios (Goldsmith and O'Regan 2005). Along with companies specialising in post-production services, such as the Universal Production Partners (UPP) Prague, they give the cluster a competitive edge.

2.4 Conclusions

In the debate on the creative city attention is paid to the fact that any kind of policy relating to the creative sector should focus on specific resources characteristic of individual cities, and in particular on a network of cooperation among entrepreneurs and on a flexible local labour market. Networks allow creative cooperation or competition, an exchange of ideas and the transfer of technology between entities, while a flexible labour market is crucial for occupational mobility and therefore the possibility of frequent changes of jobs or simultaneous work on several projects. Equally important are a city's infrastructure and social climate.

Although many of those resources are fully developed only in large cities like Paris, London or New York, also smaller urban areas can be creative. There are many creative small and medium-sized towns specialising in certain fields. Specialisation can therefore be a way of building a creative city. This way, however, is not the same for all urban areas. As we tried to show in this chapter, specialisation can have a different character. Crucial for specialisation based on a city's creative potential are its local cultural resources and the development path it has followed. Otherwise specialisation can result from the specific nature of a creative industry, and in the case of large cities specialising in several areas and combining the benefits of specialisation and diversification can help them achieve a competitive advantage.

Concentration and diversity are among the most important properties of cities that boost the chances of the appearance of creativity, which in turn enhances the development of a creative economy in a city. The growth of a creative city is largely stimulated by the presence of a certain number of skilled and creative people. However, as Hospers states (2003: 149), "concentration is not so much a matter of the number of people but rather of the density of interaction". This concentration creates conditions favourable to communication and interaction. Therefore, a concentration of people in one place leads to an intensification of contacts and meetings of individuals, which makes an exchange of new ideas and innovations quicker and more efficient. Such places are capable of causing the city to bloom by creating an atmosphere of collaboration, specialisation and innovation (Hospers 2003). Diversity, another factor enhancing urban creativity, should be understood in the broadest sense possible, i.e. as the social, economic and physical diversity of the city's elements. From the economic point of view, the most crucial is the diversity of its economic structure, both in terms of the labour market (availability of workers and a variety of skills) and the types of activity conducted. It is generally assumed that diversity is connected with the size of a city. However, several studies have shown the impossibility of defining unambiguously any relationship between the size of a city and the diversity of its economic structure (cf. e.g. Davies and Donoghue 1993; O'Donoghue and Townshend 2005). In fact, it is only big cities (usually with more than one million inhabitants) that display a fairly clear diversity. Smaller cities are usually a mixture of specialised or diversified centres. As Scott (2004: 477) points out, "where appropriate local assets are available and effectively mobilised, even quite small centres can maintain a lasting presence in the global cultural economy". In this way specialisation can become a driver of development for small cities (as demonstrated by the case of Cremona), medium-sized ones (the example of Leipzig), and large urban areas (such as Prague). All of them use both specialisation and diversity to maintain their competitive edge.

Acknowledgement This research was supported financially from the means of the Polish National Science Centre granted by decision no. DEC-2011/03/D/HS4/01662.

References

Arrow KJ (1962) The economic implications of learning by doing. Rev Econ Stud 29(3):155–173

Asheim B, Coenen L, Vang J (2007) Face-to-face, buzz, and knowledge bases: sociospatial implications for learning, innovation, and innovation policy. Environ Plan C Govern Policy 25(5):655–670

Aydalot P (1986) Milieux innovateurs en Europe. GREMI, Paris

Bathelt H (2002) The re-emergence of a media-industry cluster in Leipzig. Eur Plan Stud 10 (1):583–611

Bathelt H, Malmberg A, Maskell P (2004) Clusters and knowledge: local buzz, global pipelines and the process of knowledge creation. Prog Hum Geogr 28(1):31–56

Bertacchini EE, Borrione P (2013) The geography of the Italian creative economy: the special role of the design and craft-based industries. Reg Stud 47(2):135–147

Bontje M, Musterd S, Kovács Z, Murie A (2011) Pathways toward European creative-knowledge city-regions. Urban Geogr 32(1):80–104
Borrás S, Tsagdis D (2008) Cluster policies in Europe: firms, institutions and governance. Edward Elgar, Cheltenham
Boschma R, Kloosterman RC (2005) Further learning from clusters. In: Boschma R, Kloosterman RC (eds) Learning from clusters: a critical assessment from an economic-geographical perspective. Springer, Berlin, pp 391–405
Camagni RP (1995) The concept of innovative milieu and its relevance for public policies in European lagging regions. Pap Reg Sci 74(4):317–340
Chapain C, Cooke P, De Propris L, MacNeill S, Mateos-Garcia J (2010) Creative clusters and innovation: putting creativity on the map. Nesta, London
Chapain C, Lange B, Stachowiak K (2013) Can medium-sized cities become creative cities? Discussing the cases of three Central and Western European cities. Rozwój Regionalny i Polityka Regionalna 24:99–127
Cooke P, Lazzeretti L (2008) Creative cities, cultural clusters and local economic development. In Edward Elgar, Cheltenham
Currid E, Williams S (2010) The geography of buzz: art, culture and the social milieu in Los Angeles and New York. J Econ Geogr 10(3):423–451
Cusinato A, Philippopoulos-Mihalopoulos A (2016) Knowledge-creating milieus: firms, cities and territories. Springer, Heidelberg
Dahlström M, Hermelin B (2007) Creative industries, spatiality and flexibility: the example of film production. Norsk Geografisk Tidsskrift – Norwegian J Geogr 61(3):111–121
Davies WKD, Donoghue DP (1993) Economic diversification and group stability in an urban system: the case of Canada, 1951–86. Urban Stud 30(7):1165–1186
Deeg R (2001) Institutional change and the uses and limits of path dependency: the case of German finance. MPIFG, Köln
Dioli I, Rizzi P (2015) Place identity, music tourism and heritage: the case of the cultural district of the province of Cremona. In: Regional studies association annual conference, Piacenza, Italy
Duranton G, Puga D (2000) Diversity and specialisation in cities: why, where and when does it matter? Urban Stud 37(3):533–555
Evans G (2009) From cultural quarters to creative clusters: creative spaces in the new city economy. In: Legner M (ed) The sustainability and development of cultural quarters: international perspectives. Institute of Urban History, Stockholm, pp 32–59
Farhauer O, Kröll A (2012) Diversified specialisation: going one step beyond regional economics' specialisation-diversification concept. Jahrb Reg 32(1):63–84
Flew T (2013) Global creative industries. Polity Press, Cambridge
Florida R (2002a) The rise of the creative class. Basic Books, New York
Florida R (2002b) The rise of the creative class: and how it's transforming work, leisure, community and everyday life. Basic Books, New York
Florida R (2005) Cities and the creative class. Routledge, New York
Fuchs G (2005) Rethinking regional innovation and change: path dependency or regional breakthrough? Springer, New York
Goldsmith B, O'Regan T (2005) The film studio: film production in the global economy. Rowman & Littlefield, Lanham
Grabher G (2002) The project ecology of advertising: tasks, talents and teams. Reg Stud 36 (3):245–262
Gugu S, Dal Molin M (2016) Collaborative local cultural governance: what works? The case of cultural districts in Italy. Adm Soc 48(2):237–262
Gugu S, Mussinelli EG, Tartaglia A (2014) From creative cities to creative territories: Lombardy's cultural district program. In: Wiktor-Mach D and Radwański P (ed.) Procedeengs: The idea of creative city/The urban policy debate, Cracow, 17–18 Oct 2013. European Scientific Institute, Cracow, 119-135
Hall P (1998) Cities in civilization. Pantheon Books, New York

Hall P (2004) Creativity, culture, knowledge and the city. Built Environ 30(3):256–258
Hausner J (1995) Strategic choice and path-dependency in post-socialism: institutional dynamics in the transformation process. Edward Elgar, Aldershot
Hospers GJ (2003) Creative cities: breeding places in the knowledge economy. Knowledge, Technology and Policy 16(3):143–162
Jacobs J (1969) The economy of cities. Random House, New York
Johnson-Yale C (2008) "So-called runaway film production": Countering Hollywood's outsourcing narrative in the Canadian press. Crit Stud Media Commun 25(2):113–134
Karlsson C, Picard RG (2011a) Media clusters: spatial agglomeration and content capabilities. Edward Elgar, Cheltenham
Karlsson C, Picard RG (2011b) Media clusters: what makes them unique? In: Karlsson C, Picard RG (eds) Media clusters: spatial agglomeration and content capabilities. Edward Elgar, Cheltenham, pp 3–29
Krätke S (2011) The creative capital of cities: interactive knowledge creation and the urbanization economies of innovation. Wiley-Blackwell, Oxford
Krätke S, Taylor PJ (2004) A world geography of global media cities. Eur Plan Stud 12 (4):459–477
Landry C (2000) The creative city: a toolkit for urban innovators. Earthscan, London
Lange B, Burdack J, Herfert G, Tahlmann R, Manz K (2007) Creative Leipzig? Pathways to creative and knowledge-based regions. AMIDSt, Amsterdam
Lazzarini SG, Miller GJ, Zenger TR (2008) Dealing with the paradox of embeddedness: the role of contracts and trust in facilitating movement out of committed relationships. Organ Sci 19 (5):709–728
Lazzeretti L, Boix R, Capone F (2008) Do creative industries cluster? Mapping creative local production systems in Italy and Spain. Ind Innov 15(5):549–567
Lorenzen M, Frederiksen L (2008) Why do cultural industries cluster? Localisation, urbanization, products and projects. In: Cooke P, Lazzeretti L (eds) Creative cities, cultural clusters and local economic development. Edward Elgar, Cheltenham, pp 155–179
Mahoney J (2000) Path dependence in historical sociology. Theory Soc 29(4):507–548
Marshall A (1920) Principles of economics. Macmillan, London
Meusburger P (2009) Milieus of creativity: the role of places, environments, and spatial contexts. In: Meusburger P, Funke J, Wunder E (eds) Milieus of creativity: an interdisciplinary approach to spatiality of creativity. Springer, Heidelberg, pp 97–153
Meusburger P, Funke J, Wunder E (2009) Milieus of creativity: an interdisciplinary approach to spatiality of creativity. Springer, Heidelberg
Mirrlees T (2013) Global entertainment media: between cultural imperialism and cultural globalization. Routledge, Oxon
Müller K, Rammer C, Trüby J (2009) The role of creative industries in industrial innovation. Innov Manag Policy Pract 11(2):148–168
Musterd S, Murie A (2010) The idea of the creative or knowledge-based city. In: Musterd S, Murie A (eds) Making competitive cities. Wiley-Blackwell, Oxford, pp 17–32
Němec M (2013) Význam kulturních a kreativních průmyslů v Evropské unii, České republice a hl. m. Praze. Útvar rozvoje hl. m. Prahy, Praha
O'Donoghue D, Townshend IJ (2005) Diversification, specialization, convergence and divergence of sectoral employment structures in the British urban system, 1991–2001. Reg Stud 39 (5):585–601
Pierson P (2000) Increasing returns, path dependence, and the study of politics. Am Polit Sci Rev 94(2):251–267
Ponzini D, Gugu S, Oppio A (2014) Is the concept of the cultural district appropriate for both analysis and policymaking? Two cases in Northern Italy. City Cult Soc 5(2):75–85
Power D (2011) Priority sector report: creative and cultural industries. European Commission, Luxembourg

Pratt AC (2008) Creative cities: the cultural industries and the creative class. Geografiska Annaler Series B: Human Geogr 90(2):107–117
Pratt AC (2011) Microclustering of the media industries in London. In: Karlsson C, Picard RG (eds) Media clusters: spatial agglomeration and content capabilities. Edward Elgar, Cheltenham, pp 120–135
Romer PM (1986) Increasing returns and long-run growth. J Polit Econ 94(5):1002–1037
Rumpel P, Slach O, Koutský J (2010) Creative industries in spatial perspective in the old industrial Moravian-Silesian Region. Ekonomie a Management 4:30–46
Sassen S (2009) The specialised differences of cities matter in today's global economy. In: Whimster S (ed) Reforming the city: responses to the global financial crisis. Forumpress, London, pp 209–236
Scott AJ (1996) The craft, fashion, and cultural-products industries of Los Angeles: competitive dynamics and policy dilemmas in a multisectoral image-producing complex. Ann Assoc Am Geogr 86(2):306–323
Scott AJ (1999) The cultural economy: geography and the creative field. Media Cult Soc 21 (6):807–817
Scott AJ (2004) Cultural-products industries and urban economic development: prospects for growth and market contestation in global context. Urban Aff Rev 39(4):461–490
Scott AJ (2006a) Creative cities: conceptual issues and policy questions. J Urban Aff 28(1):1–17
Scott AJ (2006b) Entrepreneurship, innovation and industrial development: geography and the creative field revisited. Small Bus Econ 26(1):1–24
Scott AJ (2008) Social economy of the metropolis: cognitive-cultural capitalism and the global resurgence of cities. Oxford University Press, New York
Scott AJ (2010) Cultural economy and the creative field of the city. Geografiska Annaler: Series B, Human Geogr 92(2):115–130
Scott AJ (2014) Beyond the creative city: cognitive-cultural capitalism and the new urbanism. Reg Stud 48(4):565–578
Slach O, Koutský J, Novotný J, Ženka J (2013) Creative industries in the Czech Republic: a spatial perspective. Ekonomie Manag 16(4):14–29
Staber U (2008) Network evolution in cultural industries. Ind Innov 15(5):569–578
Storper M (2013) Keys to the city: how economics, institutions, social interactions, and politics shape development. Princeton University Press, Princeton
Storper M, Scott AJ (2009) Rethinking human capital, creativity and urban growth. J Econ Geogr 9 (2):147–167
Stryjakiewicz T (2010) Location factors of the creative and knowledge-intensive industries in European metropolitan regions. Geografický Časopis 62(1):3–19
Stryjakiewicz T, Burdack J, Egedy T (2010) Institutional change and new development paths: budapest, Leipzig, Poznań, Riga and Sofia. In: Musterd S, Murie A (eds) Making competitive cities. Wiley-Blackwell, Oxford, pp 93–112
Taine H (1873) The philosophy of art. Holt & Williams, New York
Tissot L, Merger M (2001) Technological trajectories, markets, institutions. Industrialized countries, 19th – 20th centuries. In Lang, Bern
Tomczak P, Stachowiak K (2015) Location patterns and location factors in the cultural and creative industries. Quat Geochronol 34(2):7–27
Törnqvist G (1983) Creativity and the renewal of regional life. In: Buttimer A (ed) Creativity and context. Gleerup, Lund, pp 91–112
Turok I (2003) Cities, clusters and creative industries: the case of film and television in Scotland. Eur Plan Stud 11(5):549–565
Uzzi B (1997) Social structure and competition in interfirm networks: the paradox of embeddedness. Adm Sci Q 42(1):37–69
Wali KC (2010) Cremona violins: a physicist's quest for the secrets of Stradivari. World Scientific, Singapore

Walls WD, McKenzie J (2012) The changing role of Hollywood in the global movie market. J Med Econ 25(4):198–219

Wasko J, Erickson M (2008) Cross-border cultural production: economic runaway or globalization? Cambria Press, Amherst

Wu W (2005) Dynamic cities and creative clusters. In World Bank policy research. World Bank, Washington

Chapter 3
From Global to Local Creative Dynamics: The Location Patterns of Art Galleries

Jérémie Molho and Dominique Sagot-Duvauroux

3.1 Introduction

The geography of the global art market[1] has undergone dramatic changes in the last two decades (Velthuis and Curioni 2015). Chinese auction sales, which rose from a negligible amount to about one third of the global art sales, offer the most striking evidence of this shift (Art Price 2015). From Paris to New York and from New York to Hong Kong, the main actors of the global art market, such as art galleries, art fairs, auction houses, are highly mobile. But while most of the attention is focused on major auction houses and galleries, less is mentioned about the bottom of the iceberg, which consists of thousands of art galleries that not only trade in artwork but also construct artistic values, by promoting artists. The aim of this chapter is to describe and explain the spatial concentration of art galleries on the scale of regions, cities, and districts. What factors cause certain territories to attract a large number of art galleries? Why do art galleries cluster?

Different kinds of art galleries can be identified. Some just sell artwork without signing any contract with the artists or promoting them. Others galleries represent artists and are in charge of the construction of their reputation (Moureau and Sagot-Duvauroux 2016). Art Galleries are polarized in certain areas such as North

---

[1]Our definition of the art market comprises both fine arts, including paintings, sculpture or drawings, and contemporary art, which covers a wider range of artwork such as artistic installations, video art or performances.

J. Molho (✉)
Robert Schuman Centre for Advanced Studies, European University Institute,
Via Boccaccio 121, 50133 Florence, Italy
e-mail: jeremie.molho@eui.eu

D. Sagot-Duvauroux
GRANEM, Angers University, 13 allée François Mitterrand, 49046 Angers, cedex 01, France
e-mail: dominique.sagot@univ-angers.fr

© Springer International Publishing AG 2017
C. Chapain, T. Stryjakiewicz (eds.), *Creative Industries in Europe*,
DOI 10.1007/978-3-319-56497-5_3

America, Europe, or East Asia. Within these regions, they concentrate in a few cities and within those cities, they agglomerate in a couple of neighborhoods. In order to explain this multi-scalar polarization, two approaches need to be combined. The first one is based on the analysis of the rules of the art market, a market which is organized in "worlds" (Becker 1982) that share common values and operate in networks to convert an aesthetic value into an economic value and a price. This perspective argues that geographical proximity plays a major role in the emergence of local art scenes and in the construction of the economic value of artworks. The second approach mobilizes economic geography to understand how some places can rise up as hubs of art trade, given their social and economic characteristics.

The first section presents the key lessons that can be learned from the academic literature on the art market and on economic agglomeration in order to shed light on the location patterns of art galleries. These two fields have a tendency to be disconnected in academic literature. The second section introduces our methodological approach The third section describes the geographic distribution of galleries in the world on a global and a local scale. We go on to propose an explanatory framework for this geographic distribution. In the discussion of the chapter, we introduce a distinction between "hubs" and "scenes" and argue that successful territories are those that combine these two functions into art market nexuses.

3.2 Literature Review

We draw on two types of literature. On the one hand, a first body of literature describes and analyzes the organizational features, economic mechanisms and spatial patterns of the art market. Some of these works may refer to space but they are not dedicated to explaining the spatial patterns of the art market and pay little attention to its multi-scalar dynamics. On the other hand, the second body of literature is concerned with economic geography, which provides theories on the location of firms, rarely applied to the art market.

3.2.1 Explaining the Organization of the Art Market

First, several works analyze the rules of the art market without referring to its spatial concentration. Art sociologists have developed theories on the construction of the artist's reputation. For example Howard Becker (1982) describes the process of reputation construction in different "art worlds." Each art world has its own price setting mechanism and is weakly connected with one another (see also Moureau and Sagot-Duvauroux 2012). Heinich (1999) distinguishes three "artistic genres" which refer to specific aesthetic conventions: the classical, the modern, and the contemporary. Moulin (1992) shows that the value of artwork results from interactions among different actors involved in the art market, not only galleries and curators but also collectors and critics. From an economic point of view, Moureau and Sagot-Duvauroux (2010) underline the consequences of uncertainty regarding the quality of contemporary art work on the market rules. They show that in the face of this uncertainty, poorly informed collectors may be rationally led to adopt copy-cat forms of behavior which consist in following the opinion of a few agents who are believed to be "in the know" concerning the value of the works (dealers, curators or collectors), involving informational cascades which can lead to speculation on some artists. In the same way, Velthuis (2005) explains the complex mechanism of art value building. Bonus and Ronte (1997) defend the idea that this process is an example of "path dependent institutional change." They show how, over and above the talent of the artist, recognition depends on small-scale events and/or luck (meeting with a curator, for example). Some papers propose gallery typologies that distinguish "promotion galleries" which primarily support innovative art, as part of a complex network of art legitimation and have contracts with artists, from "point-of-sale" galleries which only host exhibitions and sales (Moureau and Sagot-Duvauroux 2012). All these works underline the role of the networks in the process of value building but without digging into the role of spatial proximity in this process.

In parallel, numerous economic papers investigate the rate of return of art work investments (Baumol 1986; Burton and Jacobsen 1999; Ashfenlter and Kathryn 2003; Mandell 2009). Some of them show that the selling price depends on location. Thus, De la Barre et al. (1994) demonstrate that the same type of work (great masters, or other painters) fetched higher prices in New York than in London, and higher in London than in Paris.

While this literature review is necessary in order to start contemplating the geographic organization of art galleries, one needs to dig deeper with the use of spatial dynamics analysis. On a global scale, some of the literature discusses the formation of art centers through various angles: Quemin (2006) uses rankings such as the Kunstkompass or the Artreview Power List[2] to give evidence of a center-periphery dynamic within the international art market. His analysis reveals the domination of a minority of Western countries, but rarely investigates the infra-national scale. Based on auction data, Hwang (2011) analyzes the role of factors such as GDP growth or the weight of the insurance and the financial sector on the rise of a city as an international art center. Castelnuovo and Ginzburg's (1981) sociohistorical analysis goes back to the Renaissance to stress the role of art thinkers in the reputation of art centers. They show their capacity to link artistic innovations with certain places. Furthermore, research ranging from historical to geographical and sociological sources investigate the rise of the art market in cities

[2]The Art Compass is a ranking of artists performed initially by the German magazine Capital. The ranking is based on their presence in major art institutions and magazines. To each location or magazine, is allocated a number of points (e.g., MOMA 400 points) and the artists in these places or reviews accumulate these points and can thus be compared to others. Art Review publishes an annual ranking of the 100 most influent personalities of the market (Power 100). This ranking is made by a sample of experts.

such as Paris (White and White 1965), New York (Crane 1989), or London (While 2003). Usually these works go beyond the mere study of art galleries and also look at the role of art institutions and at the organizational dynamics of major artistic movements. Some recent papers have been investigating emerging art markets, such as Brazil (Brandellero 2015), India (Ithurbide 2012), or China (Joy and Sherry 2004), and pointed at the dynamics of some specific cities.

On a local scale, monographic research analyzes the emergence and the evolution of art gallery districts in major art centers like Paris (De Maupeou and Saint-Raymond 2012), New York (Molotch and Treskon 2009), Seoul (Kim 2007), and Barcelona (Rius-Ulldemolins 2012). Other research compares the spatial dynamics of various actors of the art world, in case studies such as Berlin (Boichot 2014), Brussels (Debroux 2012) or Paris (Waellisch 2010). A few articles adopt a multi-scalar approach such as Verlaine (2008) on Paris, Ithurbide (2014) on Mumbai, or Molho (2016) on Istanbul. On the whole, these works illustrate the complex interaction between the evolution of the art market and its changing urban patterns.

3.2.2 The Spatial Organization of Firms

The economic geography tools are scarcely mobilized to explain the spatial organization of the art market. Many articles seek to understand firms' location choices (see for example Gordon and McCann 2000; Bathelt and Glücker 2003; Karlsson 2008; Beccatini et al. 2009). This literature has been used to understand the onset and development of creative clusters (Cooke and Lazzeretti 2008; Chesnel et al. 2013). According to Gordon and McCann (2000), the economics of industrial clusters consist of three complementary analytical approaches.

- Following Marshall (1890), the spatial economics approach identifies centrifugal and centripetal forces that drive companies to agglomerate (Krugman 1991; Fujita et al. 1999). A territory is characterized by a number of natural or built-over-time resources, sometimes resulting from historical accidents. These resources predispose certain specializations because they give territories a competitive advantage. In the cultural field, heritage can play a key role in the emergence of a tourism cluster. Meanwhile, cities facing industrial decline have often used cultural investments as an opportunity to boost their growth and regenerate industrial districts.
- The second approach draws on Saxenian (1996) or Porter (1998), and seeks to answer the following questions: To what extent do inter-organizational networks improve the efficiency of a cluster and its constituent organizations? How do clusters strengthen the local economy? What should the role of public authorities be in these organizations? These analyses describe interactions within a complex network of private and public organizations, they highlight formal cooperation such as partnership contracts as well as informal cooperation derived from built-up trust and the sharing of common values.

– The last approach deals with the role of interpersonal relationships in the development of clusters (Granovetter 1973, 1985). Cooperation between organizations arises from decision-makers' personal relationship, leading to an approach which investigates social embeddedness of firms of a certain territory. It highlights the role of leaders who have the capacity to set up collaborative projects, and of third places that enable encounters, the matching of skills, and the launching of collective initiatives (Cohendet et al. 2010).

Drawing on the existing literature, this chapter argues that the location patterns of art galleries derives from two mechanisms: On the one hand, physical proximity and network dynamics participate in the constitution of art worlds characterized by common aesthetic values and this contributes to the construction of the economic value of artworks. On the other hand, the art market is more likely to thrive in certain places, given their physical and socio-economic characteristics.

3.3 Methodology

The main challenge for analyzing the distribution of art galleries, whether on a global or city scale, is the absence of an exhaustive and standardized database, gathering data about existing galleries and their location. In order to go beyond this methodological obstacle, we had to combine various databases indicating the location of art galleries. On a global scale, we used the Artfacts.net data to give a general view of the distribution of art galleries worldwide and to visualize the number of art galleries per city. Artfacts.net is an online archive of art exhibitions going back to 1860. It provides lists of art organizations per country and per city, sorted by types. On a city scale, we collected data on the precise locations of art galleries within twenty cities in the world, chosen according to the number of galleries, based on the Arftfacts.net data, and with a balance between developed and developing countries. In each of these cities we compiled data from various specialised sources, indicating the precise address of each gallery. We used sources ranging from specialized guides to art gallery associations.[3] Given the possible bias associated with each individual sources, we tried to confront several sources for each city, whenever it was available.

After gathering information on galleries' locations, we compiled additional data regarding poles concentrating large numbers of galleries, in order to understand which factors were relevant to explain such concentrations. Specialized reports such as those of Artprice or TEFAF (McAndrew 2015) provide comparative data on various countries, which are useful for understanding the polarization of art

[3]For instance, in Paris, the Association of Parisian Art Galleries provides a mapping of several art gallery districts (http://www.associationdesgaleries.org/). For the Singapore case, we used the Singapore Art Gallery Guide (http://sagg.info/), and in Istanbul, the magazine *Exhibist* (http://www.exhibist.com/).

galleries: number of high net-worth individuals (HNWI), average of wealth allocated to art and auction sales results. For each city, we collected additional information from professional reports, academic works, or specialized press articles.

In this chapter, we chose to focus on three cities, New York, Istanbul and Singapore. This choice was based on the availability of data for analyzing galleries' location patterns, and on the relevance of these cities as dynamic art centers. First, we could mobilize many sources of information on these three cities. For the New York case, many sources indicate art galleries locations. We especially drew on the data compiled by Earl Bateman and available on his website,[4] which were congruent with data found in other sources such as Artcollecting.com. As for explanatory factors, we relied on numerous academic works, both on New York as a whole (Guilbaud 1989; Cohen-Solal 2009; Zukin 1989; Currid 2007) or specifically on art gallery districts (Szántó 2003; Molotch and Treskon 2009; Halle and Tiso 2008). For Istanbul and Singapore, we used data collected during field trips, where we conducted respectively seventy-one and eleven interviews with key actors of both cities' art market.

Secondly, these three cities are especially relevant to the analysis of the location patterns of art galleries. New York is considered as the main center of the international art market. It has been a well-established market for decades and has had art gallery districts since the nineteenth century enabling us to conduct a longitudinal analysis of art galleries location patterns. As for Istanbul and Singapore, within both cities, in the last two decades, there has seen a large increase in the number of art galleries and they are both positioned as centers of emerging regional art markets. On the one hand, Singapore aims to become a hub for Southeast Asian Art and on the other, Istanbul's rise as an art center mainly derives from its dynamic art scene. These different processes give rise to different clustering dynamics.

In order to analyze the data, we start by providing a mapping of art galleries along with a descriptive analysis of location patterns and we highlight the characteristics of the territories that tend to concentrate art galleries. Based on these empirical insights, we propose an explanatory framework arguing that a series of forces and dynamics favor the concentration of art galleries in certain areas. To provide evidence for the effects of these forces, we draw on emblematic examples.

3.4 Descriptive Analysis of the Distribution of Art Galleries

This part aims to provide a descriptive overview of the distribution of art galleries on two scales. We start by presenting the data on a global scale, showing the concentration of galleries in a small number of countries and cities. Then we go on with describing the distribution of art galleries within an established art market center, New York and two emerging ones, Singapore and Istanbul.

[4]http://www.nycartspaces.com/

3.4.1 Global Scale

Based on the data of Aftfacts.net, we compiled the number of galleries per city in the world as well as the percentage of the total art galleries of the database per country. The map below illustrates the concentration of art galleries in a few countries and in a small number of cities. The United States (18%) and Germany (17%) gather more than one third of the total number of art galleries worldwide. Western European countries such as France, the United Kingdom (7%), Italy (6%), Spain, and Switzerland (4%) make up a quarter of the total. The BRICS countries, who arguably play an increasing role in the international art market (Velthuis and Curioni 2015) gather only a small proportion of world's art galleries: 2.2% in China, 0.9% in Brazil, 0.6% in Russia, and 0.5% both in India and South Africa. The analysis of the distribution of art galleries confirms Quemin's (2006) findings based on artists rankings[5]: the United States and West European countries appear to dominate the international art market. While many observers have predicted the rise of emerging countries in the global art market, the distribution of art galleries suggests a continuity in the uneven development of the international art market. The art market is not only concentrated within a few countries, but also within a few cities. New York attracts by far the largest number of galleries (1251), followed by three European cities: Paris (897), London (845), and Berlin (820). Although thirteen out of the twenty cities, that gather the most art galleries, are located in Western Europe; some art centers stand out in emerging countries, especially in Asia, with Beijing (255), Seoul (237), Shanghai (117), Hong Kong (90), Taipei (86) or Singapore (84), as well as in Latin America with Sao Paolo (80) and Buenos Aires (194) (Fig. 3.1).

The distribution of art galleries on a global scale reveals the intercity network structures of national and regional art markets. In many countries and regions, the art market tends to concentrate in one city. This is the case in France, with Paris or in the United Kingdom with London, but also in emerging countries such as Argentina, Turkey, Korea, or Mexico. In other cases, two or more differentially positioned cities concentrate the bulk of the country's art galleries. It is the case in Germany with Berlin as a dynamic creative contemporary art capital, and Dusseldorf and Cologne as wealthy art market centers. Some other West European countries have two or more art centers, such as Switzerland, Italy, Belgium, and Spain. In the United States, in spite of clear hierarchy in terms of the number of galleries, cities position differently. Whereas New York is the main art market capital, Chicago has historically been developing a distinctive style, with for example, the Imagists in the mid-sixties (Warren and Abell 1996). Los Angeles emerged progressively as a distinctive art center with the creation of art schools and art institutions after World War 2 (Szántó 2003), and Miami was recently positioned as a gateway to Latin American art markets (Thompson 2011). The three

[5]Quemin, A. (2006). Globalization and Mixing in the Visual Arts: An Empirical Survey of 'High Culture' and Globalization. *International Sociology, 21*(4), 522–550.

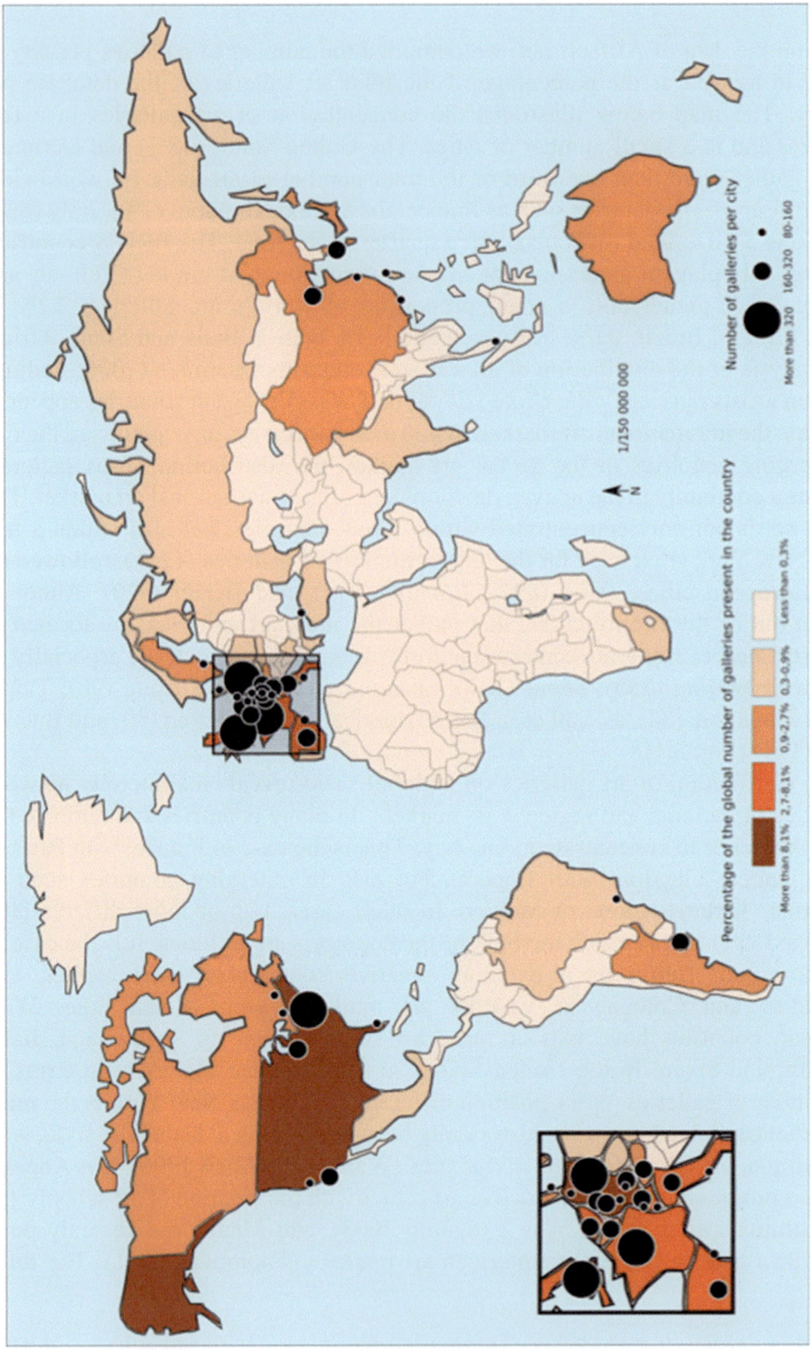

Fig. 3.1 The distribution of art galleries on a global scale

main Chinese cities also hold distinguished positioning on the art market: Beijing concentrates a large population of artists, Shanghai is more design-oriented, and Hong Kong is presented as an intermediary between the Western and the Chinese art markets (Joy and Sherry 2004). This combination of competition and complementarity across a few cities seems to be quite common in emerging art markets. It has been shown in BRICS countries such as India with New Delhi and Mumbai (Ithurbide 2012), and Brazil with Sao Paulo and Rio de Janeiro (Brandellero 2015). This process can also go beyond state borders, like in the Persian Gulf with Doha, Abu Dhabi, and Dubai that contribute in a complementary way to the rise of the so-called Middle Eastern art market (Moghadam 2012).

3.4.2 Local Scale

We have shown the polarization of art galleries happening on a global scale, we shall now go on to show similar patterns on a local scale, within cities. We present three case studies: New York, Istanbul and Singapore. Despite differences in their history, their spatial configuration, their art market, the location patterns of these cities' galleries show some similarities.

Based on the Artcollecting.com database, three quarters of all art galleries in the New York metropolis are located in Manhattan. Although the rise of real estate values has led artistic communities to move toward Brooklyn, this borough only represents one tenth of the galleries of the New York metropolis (Fig. 3.2).

Although it was strongly impacted by the 2008 financial crisis, Chelsea has remained New York's main art gallery district, with more than two hundred galleries in 2011.[6] The second pole is Upper East Side, with nearly two hundred galleries. SoHo and the 57th street come next, with almost one hundred galleries each, while Lower East Side, Brooklyn, and Dumbo gather around 50 galleries.

The urban structure of New York's art gallery sector results from a long historical trajectory. From the nineteenth century to the first half of the twentieth century, art gallery districts followed urban development and the progressive migration of affluent populations toward the North. Upper East Side and 57th street emerged as art gallery districts in the first half of the twentieth century, the former is now hosting major galleries that operate on the secondary market, like Wildenstein or Acquavella (Szántó 2003). Thanks to the presence of important cultural institutions such as the Metropolitan Museum of Art or the Solomon R. Guggenheim museum, this district is attractive for the art audience and many important collectors live there. As for the 57th street, it is a high-end shopping destination and has art galleries on upper floors. In the mid-twentieth century, as New York affirmed as an international art center, downtown districts emerged as local scenes: first Greenwich Village, then SoHo in the 1960s and the 1970s (Crane

[6]http://www.nycartspaces.com/gallery_history.php

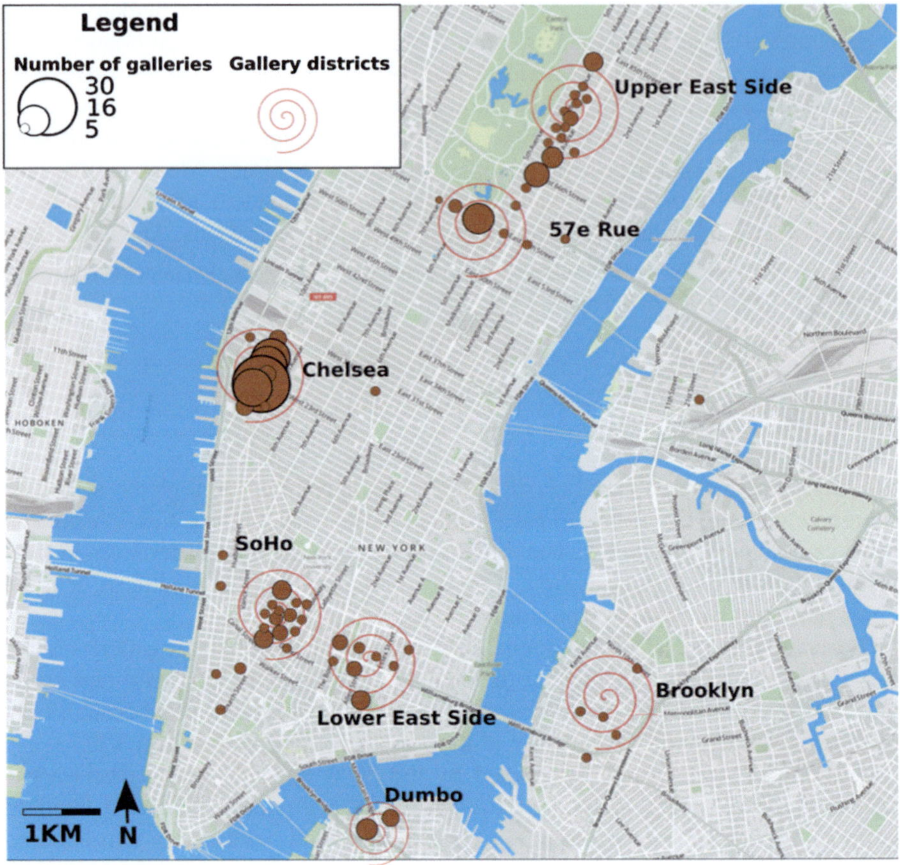

Fig. 3.2 The distribution of art galleries in New York City

1989). The latter used to be an industrial district, which was initially meant to be erased to make way for the construction of a highway. Finally, it was preserved and hosted many artists who set up numerous lofts from the 1960s onwards. The district attracted major galleries such as Leo Castelli or Paula Cooper and became the main art gallery center in the 1970s. Its decline was due to rise of real estate values and the competition of other sectors such as fashion, entertainment and tourism. While SoHo remained an art gallery district, Chelsea has become a hot spot for contemporary art (Molotch and Treskon 2009). In the 1990s, after the installation of the Dia Foundation in West Chelsea, this former meatpacking district became the new destination for contemporary art and gathered more galleries than SoHo by the turn of the twenty-first century. In parallel, Brooklyn, Dumbo, and Lower East Side, which used to be popular districts, have been attracting a rising number of galleries escaping the high rents of the main art gallery districts (Zukin and Braslow 2011).

In the case of Istanbul and Singapore, although the art market emerged more recently, we observe the formation of different art gallery districts. Two main poles concentrate the vast majority of galleries in Istanbul (Molho 2014). On the one hand, the Nişantaşı emerged in the 1980s as an art gallery district. This wealthy neighborhood hosts fancy cafés, restaurants, and luxury brands. Its galleries tend to be established and to represent generations of artists who are already recognized. This neighborhood also hosts most of Istanbul's auction houses. On the other hand the Beyoğlu district emerged in the 2000s following the rise of the Istanbul art market, and after the opening of world standard art institutions, like Istanbul Modern, SALT, Arter, and Pera Museum. Although this neighborhood has played a central role in the city's cultural history from the nineteenth century, it has degraded since the 1970s. However progressively galleries have settled in formerly degraded areas like Asmalimescit, working class neighborhoods like Tophane, or industrial districts, like Karaköy, regenerating the area (Fig. 3.3).

The location pattern of Singapore galleries is more multipolar. Whereas in Istanbul, galleries tend to cluster into axes, in Singapore, they often gather in one single building. It is the case of the Old Hill Street Police station. This colonial building founded in 1934 as police barracks was rehabilitated and hosts the Ministry of Information and Communication as well as the Ministry of Culture,

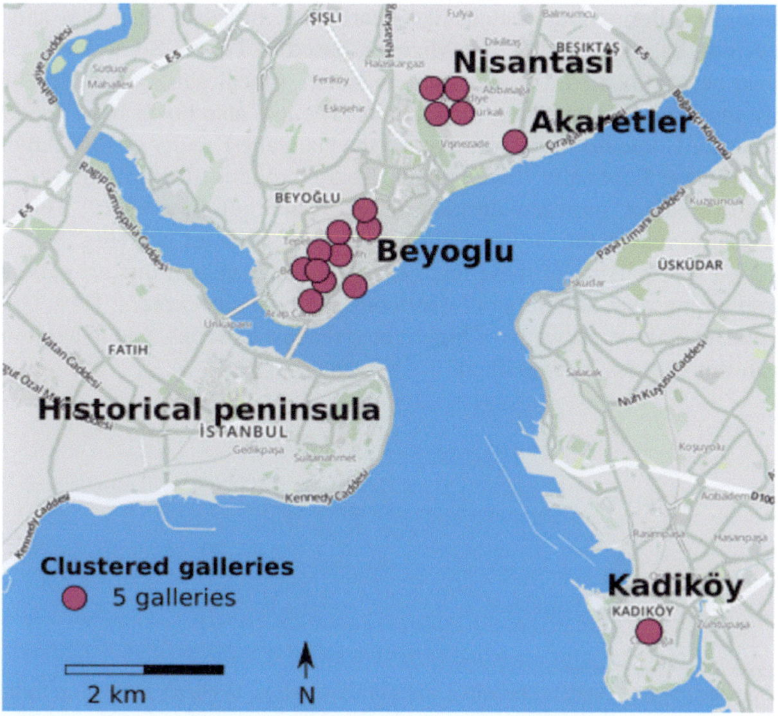

Fig. 3.3 The distribution of art galleries in Istanbul

Cooperation and Youth. The ground floor, named the "Artrium," hosts several galleries as well as cultural events. The galleries of the Raffles Hotel shopping arcades are within Singapore's oldest hotel. It was built in the nineteenth century and contains restaurants and luxury brands like Louis Vuitton. These two buildings are located in an area designated as a cultural district by the Urban Redevelopment Authority. It is the area that concentrates most art institutions such as the Singapore Art Museum, the History Museum, the Substation, as well as the two main art schools: NAFA and the Lasalle College of the Arts. Another cluster of galleries in one single building can be found in Orchard, the city's main high-end shopping area. Numerous galleries ranging from antiques to modern art are located inside shopping centers. Apart from these central gallery districts, two other districts can be pointed out. The smaller one is Tanjong Pagar Distripark, located in a logistic building owned by the Helu-Trans art transportation company,[7] which provides spaces for galleries. These galleries are quite diverse and range from aboriginal art to experimental interactive artwork or video art. Finally, the main art gallery cluster of Singapore is the Gillman Barracks. With the support of Singapore government, seventeen galleries are based in former British military barracks located on a 6.4 ha site. Thanks to a public investment of ten million dollars, 4200 m^2 were provided to galleries and 4800 m^2 were allocated to other art related organizations. Although the barracks are not directly contiguous to the main cultural and commercial centers of the city, they are relatively easily accessible. This is where most important and international galleries are located, like Pearl Lam, based in Shanghai and Hong Kong; Arndt, based in Berlin; and Mizuma, based in Tokyo. Some local galleries have also been allocated a space, such as Faust, a local gallery founded in 2006 strongly involved in supporting local artists. The site also hosts artist residencies, the Center for Contemporary Art linked to the Nanyang Faculty of Fine Arts. As an official from the Economic Development Board explains, this art gallery cluster's goal was to create a "premium destination for art in South East Asia."[8] This project takes a central role in the city's positioning as a regional art hub (Fig. 3.4).

Through the analyses of these three cases, we have shown that art galleries are polarized within a small number of districts, usually in city centers. Thus, both on a global and a local scales, we have highlighted that art galleries tend to agglomerate. How can this process be explained?

[7]This company provides both art handling and storage services, to clients such as art institutions, auction houses or art galleries. This initiative can therefore be viewed as a marketing strategy.

[8]Interview held in February 2015 with an official from EDB.

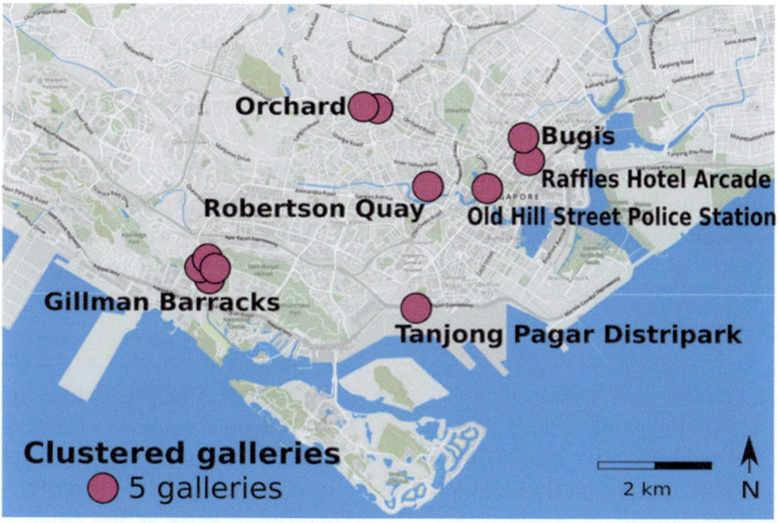

Fig. 3.4 The distribution of art galleries in Singapore

3.5 Understanding the Clustering of Art Galleries

The concentration of art galleries can be explained through two categories of factors. The first category relates to the characteristics of the territory where they are located: connectivity, wealth, urban, and cultural assets. These factors can be analyzed through what Gordon and McCann (2000) call the "pure agglomeration model." Following the spatial economic approach, it aims at understanding the agglomeration of firms in a territory, by putting forward centrifugal and centripetal forces, which repel or attract them. The second category of factors corresponds to Gordon and McCann's (2000) "industrial complex model" and derives from the strategies of local actors to improve their competitiveness on the global art market: cooperation, public support, differentiation.

3.5.1 Territorial Resources and Art Market

This section goes successively through three factors driving the concentration of art galleries: connectivity, which favors the access to potential customers; wealth, which can be allocated to art; as well as urban and cultural assets.

3.5.1.1 Connectivity

Because of the globalization of the art market, curators, collectors, and other actors of the market are continuously on the move. Art galleries looking to reach out to the international art market have to be located in cities with efficient transport infrastructures and airports. Within cities, art galleries tend to prefer central locations as they can be easily accessed from other areas of the city. An interesting exception is the case of Gagosian, the world's largest gallery in terms of gross revenue. It recently opened a branch in Le Bourget. At first glance locating in this popular northern Parisian suburb, far away from the Parisian art scene centers, may seem surprising. But Le Bourget hosts the largest private jet airport in Europe. It is therefore well located for the type of clientele base that Gagosian is targeting.

3.5.1.2 Wealth

Artistic creation and the art market have always thrived in prosperous regions. Art is a superior good, whose demand grows as revenues increase. From ancient times, the prosperity of cities such as Athens, Rome, Florence, Antwerp, and Paris has fueled artistic production and led to the rise of art trade. More recently, in the middle of the twentieth century, New York overthrew Paris as the center of the international art market as the United States affirmed as a superpower. In the 1970s, the leadership of galleries such as Leo Castelli, the Americanization of Sotheby's through the acquisition Parke-Bernet turned New York into the main international center for the creation and of the trade of artwork (Moulin 1992). During the 2000s, the emergence of the Chinese art market suggests that fast growing economies with large amount of accumulated wealth are favorable to the rise of an art market. In rich cities, artists can find financial support more easily, and galleries can develop more links with collectors.

Art galleries often choose to locate in affluent neighborhoods. Sometimes they agglomerate in areas where wealthy people live like Upper East Side in New York or Nişantaşı in Istanbul. Other art gallery districts are high end shopping districts, like the 57th street in New York, Matignon in Paris, and Orchard in Singapore. In some cases, art galleries locate near business districts. In Hong Kong, most galleries are located in Central, near the financial district, and in Dubai, some galleries are located in the Dubai International Financial Centre.[9]

3.5.1.3 Cultural Assets

Cultural amenities tend to attract art galleries. For example, in New York, in the first half of the twentieth century, when the MoMA settled in Upper East Side, it catalyzed its transformation into an art gallery district. Later Upper East Side progressively hosted many other art institutions and became known as the Museum Mile. The density of its cultural assets allows this neighborhood to retain a large

[9]Location patterns of art galleries in these cities were observed as part of the analysis of 20 cities mentioned above.

number of galleries despite its high rents. At the end of the twentieth century, the Dia Art Foundation pioneered the development of Chelsea as an art gallery district (Molotch and Treskon 2009). In Paris, the creation of the Beaubourg Art Centre gave rise to a new art gallery district in the 1970s (De Maupeou and Saint-Raymond 2012).

Emerging art market centers have been very proactive in setting up new cultural institutions. For Istanbul and Singapore, this had a quick impact on the art gallery sectors. In Istanbul many private museums and cultural institutions were created in the 2000s: Istanbul Modern, Sabancı Museum, Pera Museum, and SALT art center (Polo 2013). This gave rise to a shift in the center of gravity of galleries: within a couple of years, it moved from Nişantaşı to Beyoğlu. In 2004, the opening of the Istanbul Modern museum in a former warehouse of the Karaköy port led to the settling of new galleries in the surrounding neighborhoods. In Singapore, the government set up new cultural infrastructures such as the Singapore Art Museum, the Centre for Contemporary Art, and the National Gallery (Kong 2012). This helped the city-state attract foreign art galleries.

3.5.1.4 Urban Assets

Young galleries are often drawn to districts offering premises at lower rents. In New York, art galleries migrated toward affordable neighborhoods. Along with the evolution of contemporary artworks' dimensions, traditional premises became inadequate as larger spaces were required. In the 1960s and 1970s, derelict industrial buildings in SoHo were rented to artists and galleries (Simpson 1981). These large spaces enabled them to develop new artistic forms such as large installations and performances. But as the districts became more popular, galleries looked on for other neighborhoods. Chelsea was also attractive because it had low rents and large premises allowing some galleries to purchase property at an early stage before rents rose (Halle and Tiso 2008). The more recent move toward Lower East Side, Brooklyn, or Dumbo results from the rise of rents in established art districts. On a larger scale, in the United States, Lough (2014) argues that the high rents in cities such as Chicago, San Francisco, Seattle, and New York, constitute obstacles to the development of dynamic art scenes, whereas in cities like Los Angeles or Miami some areas have remained affordable, like Culver City in Los Angeles, and Wynwood in Miami.

This issue is crucial in emerging art markets where the values of artworks is not established and rent values are unstable. In Istanbul, when galleries settled in districts like Tophane and Karaköy, real estate started to rise, which limited the emergence of an art gallery district in this area. Kim (2007) describes a similar process in Seoul where the Cheongdam district attracted many galleries following the rise of the art market in the 1990s. Despite its relatively difficult access and the absence of cultural amenities, this district benefited from low rents and the proximity to the affluent neighborhood of Gangnam. However, in the end of the 1990s, both a crisis in the art market and the rise of rents forced half of the galleries to leave this district.

3.5.2 Actors Strategies and Clustering Logics

Beyond static characteristics that draw art galleries, local actors can build new assets in order to improve their competitiveness. Galleries can support new artistic movements and set up collective promotional events. Public and private actors can also play a major role in the clustering of art galleries.

3.5.2.1 Networks of Galleries

Art galleries' location patterns derive from clustering dynamics involving various actors of the art market such as art institutions, curators, galleries, collectors, or critics. These may deliberately collocate in order to cooperate. In the 1960s, in order to challenge the supremacy of Paris on the art market, German galleries in Cologne partnered with New York galleries. These exchanges led to the marginalization of French artists on the international stage. Recently, due to the rising pressure of auction houses, art galleries have been increasingly cooperating, through the creation of art gallery associations. They organize collective events, coordinate the dates of their opening, and develop common marketing materials. Art fairs have been mobilized as an instrument to reach a more international audience and develop networks with foreign galleries (Velthuis and Curioni 2015). The creation of an art fair can have a great impact on a local art market. When Art Basel expanded to Miami in 2002 and to Hong Kong in 2011 these cities gained a greater visibility on the international stage.

Galleries can collectively contribute to the emergence of art galleries districts. In 1970, in SoHo, the settling of the leading dealers Leo Castelli, Ileana Sonnabend, Andre Emmerich, and John Weber in the building located at 420 West Broadway participated in the image of this district as a new center for the New York art market. Likewise, in Istanbul, in 2004, Murat Pilevneli settled his gallery in the Misir Apartmani building, located in the heart of the main commercial street of Beyoğlu, Istiklal, and managed to convince other art gallery owners like Moiz Zilberman and Haldun Dostoğlu to follow him.

The collective organization of galleries can also help enhancing their collective visibility. This is why, in Paris, many art gallery district associations were founded to organize events: Art Miromesnil in Matignon, Carré Saint-Germain, and Art Saint-Germain in Saint-Germain-des-Prés. In the Belleville neighborhood, a biennial was launched. In Istanbul, when some galleries settled in the Tophane neighborhood, they created the Tophane artwalk, an event taking place every 2 months where galleries were opening at the same time.

To set up this event, they had to coordinate and elaborate common marketing materials. This event contributed to attracting an art audience to the neighborhood.

3.5.2.2 Public or Private Urban Developers

In some cases, public actors or private urban developers may adopt measures to attract art galleries in order to enhance urban value. For example, since the beginning of the 2000s, the Singapore government set up a national policy called "Renaissance" aimed at transforming the city-state into the cultural hub of Southeast Asia (Chang and Lee 2003). In order to achieve this goal, they set up several measures. They built new art infrastructures and assembled a public Southeast Asian art collection. They supported the Art Stage Singapore art fair directed by a former director of Art Basel, Lorenzo Rudolf. They supported the art gallery association of Singapore and provided tax exemption to foreign cultural companies willing to settle in Singapore. But one of the main tools of this strategy has been the setting up of an art gallery cluster in the Gillman Barracks. The Economic Development Board (EDB) of the city-state initiated this project. It consisted in the rehabilitation of former British military barracks to house art galleries. The National Arts Council took charge of the cultural animation and a Centre of Contemporary Art was set up under the authority of the Nanyang University. The EDB convinced many international galleries to settle in Singapore by promoting Singapore as a springboard to the Southeast Asian market and allowing them preferential rents.

In Istanbul, due the privatization of the cultural sector in the second half of the twentieth century, leading industrial families like Koç, Sabancı, Eczacıbaşı and Şahenk, played a key role in setting up the strategy to promote the city as an art center (Seni 2009). These families own billions-worth of holdings and are the largest supporters of the contemporary art institutions. They want Turkish artists to become successful in the international art world and by paying record prices for old Turkish Masters, like an Osman Hamdi Bey that sold for 3.5 million dollars in 2004, they took an active role in supporting Turkish art market. They also supported Turkish contemporary artists abroad, in the Sotheby's sales that took place in London in 2009 and 2010 and took directly part in the rehabilitation of Beyoğlu by purchasing buildings to set up foundations and art institutions.

3.5.2.3 Strategic Differentiation

The location of art galleries also results from strategies of aesthetic differentiation. Art gallery districts are often linked to a specific art world or genre. Galleries located in centers of historical and touristic cities often promote figurative forms of art that are distinct from contemporary art (De Vries et al. 2011). Contemporary art galleries prefer settling in unconventional places such as old industrial facilities as in the art world, location choices are linked to a strategy of differentiation from previous artistic movements. In Paris, after World War 2, numerous galleries settled on the left bank of the Seine River, in order to distinguish from galleries located in historical neighborhoods like Matignon (Verlaine 2008). They cooperated to promote new artistic movements such as Lyrical abstraction or New Realism. Likewise, in the 1970s, after the French state settled the Centre Pompidou in Beaubourg, several galleries such as Templon gallery and Beaubourg gallery, sought to differentiate themselves from left bank galleries and chose to locate near this new cultural center. In the 1980s, the creation of the Bastille Opera near the old craft neighborhood of the Faubourg Saint-Antoine, was the occasion for young avant-garde galleries to express their differences and to take advantage of premises suited to contemporary creation, like for example the Durand Dessert gallery. Later, in the 1990s, thanks to public support, young galleries settled in the 13th arrondissement near the new national library. Lately, young avant-garde galleries tended to locate in the North East of Paris, in Belleville, and in La Villette.

In Istanbul, the migration of galleries from Nişantaşı to Beyoğlu corresponds to a paradigm shift in the local art world (Molho and Birkan 2017). Whereas galleries in Nişantaşı tend to focus on Turkish modern art, exhibiting paintings and sculpture, galleries that settled in Beyoğlu adopted more contemporary aesthetics. They also exhibit performance, installations, and photography. Their approach tends to be more conceptual and experimental and they are more involved in international art networks. They are closer to what Heinich (2014) designates as a "contemporary art paradigm." Therefore indicating that the choice of location is also used to express a strategic and aesthetic positioning.

3.6 Discussion

This chapter introduced a multi-scalar approach to describe and explain the processes of polarization of the global art market. We argue that the national segmentation that observers tend to adopt is an obstacle to the understanding of the geographic dynamics of the art market. On a global scale, we show that art galleries tend to concentrate in a few cities within the North American and West European regions. The globalization of the art market and the rise of auction sales in emerging countries such as China or India did not lead to a more equal distribution of art galleries worldwide. We show that the interurban patterns of gallery agglomeration vary depending on countries and regions. On a local scale, we highlight the emergence of art gallery districts in New York, Istanbul and Singapore showing the different urban and artistic characteristics of these art gallery districts. Based on these observations, we propose an explanatory framework for the location of art galleries both on a global and a local scales and identify static locational factors such as connectivity, wealth, urban, and cultural assets. We went on to point out clustering processes: the cooperation among galleries and the involvement of other actors of the art market, the support from public or private urban developers, and the strategic differentiation of different artistic scenes.

Thus, we provide a simple explanatory framework to understand why galleries are clustered in certain areas. But this result calls for a further spatial analysis of the art market. A next step in this endeavor would require differentiating territories according to their function. Indeed, although we focus on art galleries, defined broadly as art market intermediary, economic and sociological research tend to differentiate galleries according to their functions. While some galleries are only trading artwork, promotion galleries invest in artists' careers and contribute to the construction of artistic values. Some galleries are more production-oriented and support artists by providing them with an exhibition space and by setting up artistic projects. These three types of galleries have differentiated networks within the art world (Moureau and Sagot-Duvauroux 2012). A further analysis of the spatial distribution of the global art market would differentiate territories according to their function on the art market. This analysis could study galleries' location behaviors according to their place in the value chain of the art market. Based on our investigation, we can propose a preliminary conceptualization of these differentiated territories. We can identify three types of places: art scenes, art trade hubs, and art market nexuses. First, cities or districts that have a capacity to attract artists and promote new artistic styles can be called "scenes." Second, cities or districts that offer opportunities of exchange and market development due to their financial power and their position as nodes between different market pools, can be called "hubs." Third, places that combine both attributes and therefore have the capacity to transform local production and innovation into a global market can be called art market "nexuses."

This functional specialization is not new: it was pointed out by historians in the seventeenth century in the Low Countries, when cities like Antwerp had a large concentration of painters while Amsterdam hosted auction houses and was a commercial platform (De Marchi and Miegroet 1994). But in the wake of the twenty-first century, the rise of new art market regions has translated into the rise of new hub cities such as Miami for Latin America, Dubai for the Middle East, Hong Kong for East Asia, and Singapore for South East Asia. These hub cities lack an "artistic hinterland," but they are airports and financial hubs of an emerging region. They host global art fairs and international auctions and they set up proactive strategies to develop their local art market. On the other hand, we have witnessed the emergence of scene cities such as Sao Paulo, Istanbul, Seoul, and Beijing. These cities have the capacity to attract numerous artists and host dynamic art institutions. However, their art markets tend to be more nationally oriented than those of hub cities. Finally, the reason why cities such as New York and London are considered dominant art centers is their ability to both attract art trade, and to nurture innovative art movements. They are global art market nexuses.

Within cities, districts can also be distinguished according to their function in the art market. Some districts are mainly commercial hubs, such as Upper East Side in New York, Matignon in Paris, and Nişantaşı in Istanbul. Other districts are scenes, like Kadiköy in Istanbul, Belleville in Paris, as well as Brooklyn in New York. Scenes attract fewer commercial galleries but they have large artistic communities, cooperative galleries, and they host many artistic projects. Nevertheless, the most recognized art gallery districts are those that are able to combine these two functions and constitute art market nexuses. For instance, in the 1970s, SoHo both had a high concentration of artists and the most influential commercial galleries.

The analysis of the spatial configuration of the different functions of the art market would give further insight on the location patterns of art galleries that we presented in this chapter.

References

Art Price (2015) The Art market in 2014. http://fr.artprice.com/
Ashfenlter O, Kathryn G (2003) Auctions and the price of art. J Econ Lit 41(3):763–787
Barre de la N, Docclo S, Ginsburgh V (1994) Returns of impressionist, modern and contemporary European paintings 1962–1991. Annales d'Economie et de Statistique 35, July–September
Bathelt H, Glücker J (2003) Towards a relational economic geography. J Econ Geogr 3:117–144
Baumol WJ (1986) Unnatural value: or art as a floating crap game. Am Econ Rev 76(2):10–14
Beccatini G, Bellandi M, De Propris L (2009) A handbook of industrial districts. Edward Elgar, Cheltenham/Northampton
Becker HS (1982) Art worlds. University of California Press, Berkeley/Los Angeles/London
Boichot C (2014) Les quartiers artistiques: territoires en construction. Regards croisés sur Montreuil (Î}\^{I}{le-de-France) et Neukölln (Berlin). In: Annales de Géographie, vol 698, No. 4. Armand Colin, Paris, pp 1088–1111
Bonus H, Ronte D (1997) Credibility and economic value in the visual arts. J Cult Econ 21:103–118
Brandellero A (2015) The emergence of a market for art. In: Brazil. Cosmopolitan canvases: the globalization of markets for contemporary art, 215
Burton BJ, Jacobsen JP (1999) Measuring returns on investments in collectibles. J Econ Perspect 13(4):193–212
Castelnuovo E, Ginzburg C (1981) Domination symbolique et géographie artistique [dans l'histoire de l'art italien]. Actes de la Recherche en Sciences Sociales 40(1):51–72
Chang TC, Lee WK (2003) Renaissance city Singapore: a study of arts spaces. Area 35 (2):128–141
Chesnel S, Molho J, Morteau H, Raimbeau F (2013) Les clusters ou districts industriels du domaine culturel et médiatique: revue du savoir économique et questionnement. Ministère de la culture et de la communication
Cohendet P, Grandadam D, Simon L (2010) The anatomy of the creative city. Ind Innov 17 (1):91–111
Cohen-Solal A (2009) Leo Castelli et les siens. Gallimard, Paris
Cooke P, Lazzeretti L (eds) (2008) Creative cities, cultural clusters and local economic development. Edward Elgar, Cheltenham
Crane D (1989) The transformation of the avant-garde: the New York art world, 1940–1985. University of Chicago Press
Currid E (2007) The Warhol economy: how fashion, art, and music drive. Princeton University Press, New York
De Marchi ND, Miegroet HJV (1994) Art, value, and market practices in the Netherlands in the seventeenth century. Art Bull 76(3):451–464
De Maupeou F, Saint-Raymond L (2012) Cartographie des lieux d'exposition à Paris, de 1850 à nos jours. Artl@ s Bull 1(1):4

De Vries M, Martin B, Melin C, Moureau N, Sagot-Duvauroux D (2011) Diffusion et valorisation de l'art actuel en région, Une étude des agglomérations du Havre, de Lyon, de Montpellier, Nantes et Rouen, Culture Etudes, n°CE 2011-1, http://www.culture.gouv.fr/nav/index-stat.html

Debroux T (2012) Des artistes en ville. Géographie rétrospective des plasticiens à Bruxelles, 1833–2008. Doctoral dissertation, Université Libre de Bruxelles

Fujita M, Krugman P, Venables AJ (1999) The spatial economy: cities, regions and international trade. MIT Press, Cambridge, MA

Gordon IR, McCann P (2000) Industrial clusters: complexes, agglomeration and/or social networks? Urban Stud 37(3):513–532

Granovetter M (1973) The strength of weak ties. Am J Sociol 78:1360–1380

Granovetter M (1985) Economic action and social structure: the problem of embeddeness. Am J Sociol 91:481–510

Guilbaud S (1989) Comment New York vola l'idée de l'art moderne. Jacqueline Chambon, Nîmes

Halle D, Tiso E (2008) 11 The sociology of the new art gallery scene in Chelsea, Manhattan. In: Amariglio J, Childers JW, Cullemberg SE (eds) Sublime economy: on the intersection of art and economics. Routledge, Abingdon

Heinich N (1999) Pour en finir avec la querelle de l'art contemporain. L'Echoppe, Paris

Heinich N (2014) Le paradigme de l'art contemporain: structures d'une révolution artistique. Gallimard, Paris

Hwang H-C (2011) What constitutes an International Art Center?. SSRN 1954594 [Online]. http://papers.ssrn.com/sol3/papers.cfm?abstract_id=1954594

Ithurbide C (2012) Marché de l'art contemporain indien: territoires et réseaux en construction. Transcontinentales. Sociétés, idéologies, système mondial, (12/13)

Ithurbide C (2014) Beyond Bombay art district: reorganization of art production into a polycentric territory at metropolitan scale. Belgeo, Revue belge de géographie 3(2014)

Joy A, Sherry JF Jr (2004) Framing considerations in the prc: creating value in the contemporary chinese art market. Consum Mark Cult 7(4):307–348

Karlsson C (ed) (2008) Handbook of research on cluster theory. Edward Elgar Publishing, Cheltenham

Kim H (2007) The creative economy of urban art clusters: locational characteristics of art galleries in Seoul. J Korean Geogr Soc Vol. 42. N°2 pp. 258-279

Kong L (2012) Ambitions of a global city: arts, culture and creative economy in 'Post-Crisis' Singapore. Int J Cult Policy 18(3):279–294

Krugman P (1991) Geography and trade. MIT Press, Cambridge, MA

Lough J (2014) Art vs. Real estate. Gentrification and Urban Artistic Scenes. http://artpulsemagazine.com/art-vs-real-estate-gentrification-and-urban-artistic-scenes

Mandell BR (2009) Art as an investment and conspicuous consumption goods. Am Econ Rev 99 (4):1653–1663

Marshall A (1890) Principles of economics: an introductory volume, 8th edn. Macmillan, London, p 871

McAndrew C (2015) Art market Report 2015. TEFAF (The European Fine Art Foundation)

Moghadam A (2012) L'art est mon métier : émergence et professionnalisation du marché de l'art à Dubai. Transcontinentales. http://transcontinentales.revues.org/1312

Molho J (2014) Territorialisation d'un marché de l'art émergent: le cas d'Istanbul. Belgeo, Revue belge de géographie (3)

Molho J (2016) Les territoires du marché de l'art: le cas d'Istanbul. Doctoral dissertation, University of Angers

Molho J, Birkan I (2016) Approche croisée de la sociologie et de la géographie sur le(s) monde (s) de l'art stambouliote: apports théoriques et méthodologiques. In: Langeard C (ed) Méthodes et concepts à l'épreuve du pluralisme. Presse Universitaires de Bretagne

Molotch H, Treskon M (2009) Changing art: SoHo, Chelsea and the dynamic geography of galleries in New York City. Int J Urban Reg Res 33(2):517–541

Moulin R (1992) L'Artiste, l'Institution, le Marché. Flammarion, Paris
Moureau N, Sagot-Duvauroux D (2012) Four business models in contemporary art. Int J Art Manag (IJAM) 14, n°3, spring
Moureau N, Sagot-Duvauroux D (2016) Le Marché de l'Art Contemporain. La Découverte, Paris
Polo JF (2013) The Istanbul modern art museum: an urban regeneration project?. Eur Plan Stud. http://dx.doi.org/10.1080/09654313.2013.819074
Porter ME (1998) Cluster and the new economics of competition. Harv Bus Rev 76:77–90
Quemin A (2006) Globalization and mixing in the visual arts an empirical survey of 'High Culture' and globalization. Int Sociol 21(4):522–550
Rius Ulldemolins J (2012) Gallery districts of Barcelona: the strategic play of art dealers. J Arts Manag Law Soc 42(2):48–62
Saxenian AL (1996) Regional advantage: culture and competition in Silicon valley and route 128. Harvard University Press, Cambridge, MA
Seni N (2009) Le mécène, un acteur méconnu de la ville. Istanbul à l'heure des musées privés. Transcontinentales, Sociétés, Idéologies, Système Mondial 7:105–128
Simpson CR (1981) SoHo, the Artist in the City. University of Chicago Press, Chicago
Szántó A (2003) Hot and cool: some contrasts between the visual art worlds of New York and Los Angeles. New York/Los Angeles: Politics, Society, Culture, 393-422.
Thompson ER (2011) The growth and evolution of art Basel Miami beach. Master thesis, American University.
Velthuis O (2005) Talking prices. Princeton University Press, Princeton/Oxford
Velthuis O, Curioni SB (eds) (2015) Cosmopolitan canvases: the globalization of markets for contemporary art. Oxford University Press, Oxford
Verlaine J (2008) La tradition de l'avant-garde: les galeries d'art contemporain à Paris de la libération à la fin des années soixante. Doctoral dissertation, University Paris 1
Waellisch U (2010) Geographies of creative production: the perspective of visual artists in Paris. Doctoral dissertation, © Ulrike Waellisch
Warren L, Abell J (eds) (1996) Art in Chicago, 1945–1995. Thames & Hudson, New York
While A (2003) Locating art worlds: London and the making of Young British art. Area 35 (3):251–263
White HC, White CA (1965) Canvases and careers: institutional change in the French painting world. University of Chicago Press, Chicago
Zukin S (1989) Loft living: culture and capital in urban change. Rutgers University Press, New Brunswick
Zukin S, Braslow L (2011) The life cycle of New York's creative districts: reflections on the unanticipated consequences of unplanned cultural zones. City Cult Soc *2*(3):131–140

Chapter 4
Innovation Dynamic in the Film Industry: The Case of the Soho Cluster in London

Caroline Chapain and Krzysztof Stachowiak

4.1 Introduction

This chapter analyses, in details, the innovation dynamic of the film industry within the Soho cluster in London. In doing so, the chapter aims to add to the literature looking at innovation within the creative industries (sectoral dynamic) but also to the literature on clustering within these industries (spatial dynamic) by offering an original and unique data analysis of the innovation process, linkages and outputs within a particular creative cluster.

With the rising discourse on the contribution of the creative industries to economic growth, there has been a growing interest in the way these industries innovate and how this may spill over to other economic sectors as well as to the wider society in the last 20 years (Hellbrun 1993; Tschang 2007; Müller et al. 2009; Hartley et al. 2013; Flemming 2015). In addition, much has been said on the extent to which these industries tend to concentrate in space or cluster (Chapain et al. 2010; Power 2011; Boix et al. 2015) to benefit from both economies of scale and agglomeration. While there is an assumption that spatial clustering favour economic performance in terms of creative production and innovation by facilitating the transfer of knowledge and reducing costs, etc., few work offer a detailed analysis of both the innovation dynamic of particular clusters and their innovation performance. This chapter offers to fill part of this gap by examining the Soho film cluster.

C. Chapain (✉)
Business School, University of Birmingham, Birmingham, UK
e-mail: c.a.chapain@bham.ac.uk

K. Stachowiak
Institute of Socio-Economic Geography and Spatial Management, Adam Mickiewicz University, Poznań, Poland
e-mail: krst@amu.edu.pl

© Springer International Publishing AG 2017
C. Chapain, T. Stryjakiewicz (eds.), *Creative Industries in Europe*,
DOI 10.1007/978-3-319-56497-5_4

The film industry can be highly innovative, both producing and adapting innovations as illustrated throughout its history (Bakker 2008; Bennett et al. 2008; Salt 2009; Sito 2013). This is also one of the most highly clustered of the creative industries (Tomczak and Stachowiak 2015). Nevertheless, the industry is both highly reliant on strong local and global networks of production (e.g. Christopherson and Storper 1986; Coe 2000, 2001; Scott 2005; Davis et al. 2009; Crane 2014; Kindem 2000; Miller et al. 2005; Wasko and Erickson 2008). As such, studying the innovation dynamic and performance of a film cluster should offer interesting insight in terms of the interplay between local and global sectoral and spatial dynamics. The case study of Soho in London is interesting as the UK is amongst the six most important film producers in the world (UNESCO 2016) but is also strongly connected to the American film industry. Soho occupies a central role within the British film industry in terms of its contribution to national production and crystallises its linkages with the American film industry. Nachum and Keeble (2003a, b), Cook and Pandit (2008) and Pratt and Gornostaeva (2009) have offered some interesting discussion and study of the national and international dynamics of the film cluster in London in the past. We are adding to these by offering a more in-depth analysis of the innovation dynamic and performance of the cluster at the turn of 2010.

The chapter is structured as follows. Section 4.2 summarises some of the debate around clustering in the film industry while Sect. 4.3 offers some insights on innovation in the film sector. Section 4.4 presents our research methods and Sect. 4.5 our results and findings. The last section offers some conclusion and discusses future avenues for research.

4.2 Clustering and the Film Industry

There exists a vast literature which analyses and discusses the economic geography of the film industry and its clustering process. Given the tight economic links between the film and TV sectors, some of this literature considers these two sectors together. Davis (2008) listed 80 publications analysing film and TV clusters in Europe, North America and Australasia. Various avenues of enquiry can be distinguished amongst this literature.

There is much discussion of the evolution of the film industry over time and how changes in its economic structure have been reflected in its geographical distribution and concentration (Christopherson and Storper 1986; Scott 1998, 2000, 2002, 2004; Davis 2011, Pratt 2011). The example of Hollywood and Los Angeles and the American film industry is discussed at length due to its historical role in the development of the industry and persistent economic importance throughout the years (Scott 2005; Crane 2014). The high costs of film production and high uncertainty in demand have early on led the American industry to vertically integrate into large producing studios and distribution networks bringing together the various functions and elements of the film value chain to reduce costs and minimise risks up to the 1950s, concentrating most of the industry in Los Angeles from the 1920s onwards (Christopherson and Storper 1986; Scott 2005). Changes in the US regulation obliging the film industry to reduce its concentration, increased competition from TV production and the need to keep hold of or expand on national and foreign markets has fostered a reverse phenomenon of vertical disintegration, increased flexibility and specialisation of the industry since the 1960s. Similar phenomena have been observed in France, Denmark and India (Lorenzen 2007).

Today, film production and post-production tend to be characterised by both regular intra-firm but also inter-firm project works bringing together team of creative people with various competences and skills over short periods of time (Karlsson and Picard 2011). The assemblage of these teams is rendered easier by the availability of these people in the same location. Film clusters, like many creative clusters, are thus characterised by particular creative milieus, i.e. "a shared space and tradition in which people can learn, compare, compete and collaborate and through which ideas can be proposed, developed, disseminated and rejected" (Leadbeater and Oakley 1999: 31). These milieus are characterised by high level of interactions and information exchange within interdependent networks of producers with complementary skills and expertise and by a high degree of tacit knowledge (Leadbeater and Oakley 1999; Scott 1999; Hall 2004). The disintegration and specialisation of the film industry has thus reinforced phenomena of agglomeration and concentrations of its actors around key locations across the globe whereas globalisation has alternatively led to a more complex and scattered economic geography of its various functions across these locations (Coe 2000, 2001; Currah 2003; Coe and Jones 2005; Vang and Chaminade 2007; Crane 2014). As such, cities like Los Angeles and Paris may still concentrate all the elements of the film value chain, i.e. finance, pre-production, production, post-production, distribution and exhibition. Others like Toronto and Vancouver may only specialise in some of them (Coe and Jones 2005; Vang and Chaminade 2007).

These phenomena have generated research on the characteristics, roles and functions of different film clusters (Coe 2000, 2001; Eriksson 2011). For example, some research look into the economies of scales and industrial atmosphere supported by film clusters as well as the industrial linkages and creative milieu that they generate. Some of these analysis examine in depth the governance and structure of these clusters (in terms of lead firms, small versus large firms, etc.) and their respective hierarchy, i.e. satellite production cluster (see Storper and Harrison 1991; Coe 2000; Coe and Jones 2005; Vang and Chaminade 2007). More and more focus has also been given to the role of social networks and to the importance of local embeddedness versus global networks in the transmission of knowledge and in supporting film production processes (Nachum and Keeble 2003a, b; Mossig 2004; Johns 2010). Linked to this, there is an increasing recognition of the role that cultural and social factors and that institutional actors and local and national policies play in the development of film clusters (Blair and Rainnie 2000; Karlsson and Picard 2011; Johns 2010; Crane 2014).

In their book on media clusters (including film clusters), Karlsson and Picard (2011: 4) summarise the elements to consider when looking at film cluster:

"Clusters exist when there is a large enough concentration of media (and related) firms in a city or a metropolitan area. Several firms are normally gathered in a specific, often central location in the city, but the cluster may include other firms in the surrounding area. An important element of clusters is that the cluster firms share resources, such as labour, contract services information and knowledge, and interact directly or indirectly for mutual advantage and benefit- although the degree of interaction amongst individual members may vary widely and may even be unintentional. It should also be noted that some cluster firms may also have significant interactions with players outside the cluster or in other locations". In addition, they recommend examining the vertical and horizontal economic relationships taking place within the cluster, the power and institutional dimensions at play and the influence of external relationships and factors (ibid.).

Finally, it is important to note that while many of the research on film clusters suggest that clustering supports innovation, only a few works try to assess this dynamic in depth. This is the case of Pratt and Gornostaeva (2009) and Davis et al. (2009). Pratt and Gornostaeva (2009) offer a discursive analysis of four elements which have impacted the innovation dynamic of the film cluster in London: digitisation of the industry, change in UK regulation with the introduction of favourable tax incentives for the industry, the organisation of the UK film industry as a cottage industry with strong links with the USA and finally the role of Soho in concentrating actors since the 1920s. This contribution offers an insightful background and understanding but does not document in detail the firms innovation dynamics and interactions within the cluster. In a similar exercise, Davis et al. (2009) consider five dimensions in assessing the innovation potential of the screen media based industry cluster in Ontario: understanding the specificities and nature of the innovation process within that particular industry, assessing and facilitating the linkages, spillovers and externalities supporting innovation; taking into account trans-local external linkages; considering existing innovation policies around work and entrepreneurship and opportunities for innovation policies supporting the cluster in particular. In this chapter, we propose to add to these works by *systematically measuring how clustering really impacts the innovation performance of a film industry and assess the nature of the innovation dynamic within a particular cluster*.

4.3 Innovation in Film Industry

Film is an industry that has "merged technological innovation with creative and artistic experimentation" along its history (Bakker 2008: 1). One might even say that the history of the film industry is a history of innovations in film. The very birth of film was a consequence of the invention of the cinematograph, a motion picture film camera, in the late nineteenth century. Since then, the film industry has created and adapted many innovations of different kinds. They embraced new equipment and tools and the related new ways of using them, but also novel ways of organising the film production process. All that led to vast aesthetic innovations, the emergence of film styles and genres, or the mastering of new filming techniques, to name just a few.

It is not an easy task to characterise, categorise and represent all the innovations which have occurred in the film industry throughout its history. Like other creative industries, film innovations range from tangibles such as technology or equipment to intangibles such as aesthetic or film styles whose degree of innovation lies on high degree of subjectivity (Stoneman 2010). In addition, innovations have played various roles in the film industry: they have enhanced the production processes, have generated new cinematic effects, have served as a tool for new business strategies, etc. For example, in the 1970s, the major Hollywood studios decided to reduce the number of films they produced but at the same time undertook an intense strategy of product differentiation through constant innovation to maintain their revenues (Storper and Christopherson 1987; Aksoy and Robins 1992). As such, innovation has played a critical role in the film industry since its conception.

Table 4.1 attempts to give an overview of the main innovations in the film industry over time. Innovations are distinguished along traditional "measures" of innovation in the literature such as product, process and organisational as well as soft innovation following up on the work of Stoneman (2010) about innovation in the creative industries. In addition, each type of innovation is decomposed along the various stages of the film value chain to offer a better understanding of the innovative specialisation of different film clusters. As noted in the previous section, some film clusters may concentrate all the elements of the film value chain while others may only specialise in some of them. The occurrence of specific innovations within particular film clusters will thus likely depend on their specialisation. As such, the Los Angeles region, where all stages of the value chain are present, has induced all kinds of film innovation (Scott 2005) whereas innovations in relation to digital special effects or animation have, apart from Hollywood, mostly occurred in Silicon Valley (Sito 2013), London (Pratt and Gornostaeva 2009) or Wellington, New Zealand (Margolis et al. 2008).

The history of cinematography is filled with **product innovations** based on new or improved technologies. Perhaps, the most important product innovation within the history of cinematic technology was the coming of sound in the mid-1920s to the mid-1930s. As film historian David Cook notes, "the introduction of sound is analogous in almost every respect to the invention of cinema itself" (Cook 2004: 221). Sound has transformed many dimensions of the cinema and it has had a tremendous impact on the way filmmakers told their stories and communicated with an audience (Villarejo 2007: 64). Many subsequent product innovations, such as the introduction of colour motion picture films, have constantly been transforming the industry.

One of the latest technological breakthroughs in the film industry has been the advent of digital technologies which have impacted many areas of film production, distribution and consumption. For instance, Kehoe and Matter (2015) argue that in the newly digital era film distribution is shifting from a supply-led to a demand-led market. As such, independent distributors can now break away from the rigid

Table 4.1 Main innovations in the film industry

Innovation type	Value chain stage				
	Development and pre-production	Production	Postproduction	Distribution	Exhibition
Product (technological)		Cinematograph	Sound recording and editing equipment	Analog tape formats (e.g. VHS, Betamax)	Film projectors
		Cameras	Film sound systems	Digital tape formats (e.g. Betacam, Digital8, DV)	Theatre sound systems (surround sound)
		Lenses	Film editing equipment	Optical disc storage (e.g. DVD, Blu-ray)	Architectural innovations (movie theatres)
		Sound technologies	Specialised computer equipment	Digital encoding formats (e.g. MPEG, H.264)	Stereoscopic and 3D viewing
		Colour	Server farms	Video on demand technologies (www, app, over-the-top content, IP-BOX, Set-top box etc.)	
		Equipment supporting filming (grips, tripods, rigs, dollies, cranes etc.)	(Super) computers	Streaming technologies	
		Lightning technologies	Optical fibre (for ultra-fast bandwidth)	Telecommunication infrastructure	
		Drones	Data processing equipment		
		Motion capture			
		Special effects engineering			
		Stereoscopic and 3D film recording			

Process	Storyboard	Filming techniques	Non-linear editing systems	Cable, terrestrial and satellite distribution networks	Change in theatre architecture
	Previsualization				
	Screenwriting software	Acting techniques	Computer-generated imagery	Marketing and distribution platforms	
			Software packages	Electronic marketplaces	
			Algorithms	E-commerce portals	
			Visual effects	Online distribution	
			Editing techniques		
Organisational		Studio system (1920s onwards)	Data management systems	Patenting (Edison and his cohorts in the MPPC)	Movie theatres
	Financing (e.g. specialised banks or investment funds for film, crowdfunding)	Contracting	Data backup (archives)	On-demand Internet streaming media (e.g. Netflix, Amazon Instant Video, Hulu, Apple TV)	Concession sales in the movie theatre (selling films with popcorn and soda)
	Film policies (e.g. regional, national and international policies supporting film development, regional film offices, tax incentives)	Co-productions	Project management	Film rating system	Popcorn (introduced in 1930s)
	Film policies (e.g. regional, national and international policies supporting film development, regional film offices, tax incentives)			Strategies of distribution	
	Managing creative people in film development				
	Copyright				

(continued)

Table 4.1 (continued)

Innovation type	Value chain stage				
	Development and pre-production	Production	Postproduction	Distribution	Exhibition
Soft	Literary styles (writing genres)	Film styles	Film styles	New digital content innovations	New ways of interaction with film (3D, Imax, 5D)
		Representation	Representation		Video game as an inter-active film
		Film language	Film language		
		Means of expression	Means of expression		
		Aesthetic	Aesthetic		

Source: Stachowiak (2017)

singular value chain that dominated the industry, and adopt bespoke release strategies that are more tailored to the individual needs of each film. In reality, digital technologies are supporting the creation of new film products by effectively improving mostly existing processes and making them more efficient. It is important to note that the consequences of the digitisation of the film industry have been the object of an important debate amongst filmmakers and academics.

Indeed, Elaesser (2008) points out that the classical model of cinema has adapted itself to different media and technologies over time, adjusting to the introduction of sound as well as to other technical innovations, be it colour, widescreen, animation or digital imaging techniques. Tryon (2009) notes that digital cinema works in dialogue with, and thus oddly reaffirms, the centrality of the old-fashioned theatrical experience. Dargis and Scott (2012) show that both analog and digital technologies have advantages (and disadvantages), and serve as means of artistic expression. Regardless if a film is produced through analog or digital technology, what matters is the final experience of the film for the viewers. Similar views have been expressed by many of the top contemporary filmmakers. This is well recognised in the documentary "Side by Side" (Dargis and Scott 2012, dir. Chris Kenneally) which investigates the history, process and workflow of both digital and photochemical film creation with contribution from various renowned directors from the industry. As Scott (2012: C1) notes, "the film is not quite an exercise in pro-digital propaganda. Skeptics as well as enthusiasts are heard from, and the virtues of old-fashioned celluloid are duly praised". Perhaps it is too early to fully assess the impact of digital technologies on the film industry. Nevertheless, it is evident that they have already changed the way films are produced (Dargis and Scott 2012; OECD 2008), are distributed (Zhu 2001) and are experienced (Elsaesser 2008; Isaacs 2013; Tryon 2009) as the newly digital innovations have impacted both the distribution and exhibition of films. With the rise of digital cinema, audiences often encounter movies outside the theatre and even outside the home. As such, traditional distribution models are challenged by new media entrepreneurs and independent film makers, user generated video, film blogs, mashups, downloads and other expanding networks (see Tryon 2009: for a detailed analysis of the film industry in the age of media convergence).

Another important source of innovation has been **process innovation**. Indeed, while the new technology used to produce a new type of film camera results in a new product, i.e. camera with improved features (e.g. the Arriflex 35 camera, one of the most significant motion picture cameras of all time, introduced in 1937 and upgraded until 1964 through a series of innovative changes), new ways of using a film camera or editing film content can be considered as a process innovation. As such, regardless of whether a camera is old or new, its use in novel way can lead to new cinematic effects. It is the case in new filming methods introduced such as aerial shots, close-ups, shaky cam shots, low camera angle shots and so on. Parallel examples can be found in new editing techniques such as cross-cutting or compositing (Bennett et al. 2008). Related to this, it is important to note that digital innovations are not limited to technology or product innovation, but they also embrace the film production process and thus support process innovations. For instance new software used during film production can lead to significantly improved method of delivery of particular task within the film production process. An example of such software innovation is ALICE developed by The Moving Picture Company, based in Soho, London. ALICE stands for **A**rtificial **LI**fe **C**rowd **E**ngine. ALICE was initially developed for the film "Troy" (2004) but has since been used in over 30 films: "ALICE is a crowd simulation software and is comprised of a framework of tools to simulate crowds and involves skills such as physics modelling, artificial intelligence and rendering technologies. ALICE enables artists to manage crowd behaviour, motion clip editing and blending, and customised scripting for large groups of agents when producing a film. ALICE appears to have attracted visual effects specialists to the UK from around the globe. As there is nothing else available on the market yet, visual effects practitioners travel to the UK in order to develop their skills and learn about the latest computer animation techniques" (Oxford Economic Forecasting 2005: 22).

Organisational innovation is another type of film innovations and refers to the new ways films are produced, distributed, marketed or exhibited. Bakker (2008) argues that the invention of cinema itself was an organisational innovation in the sense that it changed the organisation of the entertainment business. It altered all stages of the entertainment industry from creation to exhibition. "Distribution and organisation within the theatre became simplified, standardised and automated, while all complexity was transferred to film production, entirely outside the theatres, in a dedicated space, in which production activities were highly specialised and co-ordinated" (Bakker 2008: 164). Organisational innovations play an important role in the early stages of film production, such as financing. For example, crowdfunding is a recent organisational innovation for the industry (Sørensen 2012). Organisation innovation also occurs throughout the film production process as managing and coordinating large team of creative people comes to the fore in the film production process. New organisational ideas have emerged in terms of creative leadership, the development of team culture or the overall management of film crews within film production (Bloore 2013). Acheson and Maule (2005) argue that the American dominance in the film industry can be explained by their flexible managerial culture and their open and innovative financial system which have allowed the American film industry to take advantage of a series of historical events and technological developments. Although the American courts have viewed the industry's contracting practices and vertical integration as instruments of monopolisation, such arrangements can be defended as efficient innovations for dealing with the risk and informational gaps of film exhibition and distribution. Organisational innovations are thus used by major studios as a part of risk reducing strategies. For instance, a greater concentration in film distribution makes it possible for studios to offer cinemas a package of films. This bundling reduces transaction costs between the distributor and cinemas, increases the efficiency of a cinema's transition from one film to another, and reduces risk from portfolio effects (Acheson and Maule 2005). Such risk reducing strategies are apparent in the production and distribution of large-budget films which are often only innovative on the surface. Under this veneer lie concepts or formulae that have been successful in the past.

Another type of organisational innovation has been related to film marketing strategy. Contrary to many other creative products, films have since their inventions been considered as branded goods. Bakker (2005) explains that, during the period 1880–1930, market integration led to many business innovations such as single-price selling, branded products marketed nationally, department stores, mail-order firms, advertising agencies, market research and distribution and delivery systems and that "The film industry shared several of these innovations, especially single prices, national branding, advertising, and market research, but the most important was probably the development of a dedicated distribution delivery system, the movie theatre". Innovations in branding techniques and new possibilities of global advertising resulted in rise of specialised film marketing at the beginning of twenty-first century (Marich 2009; Kerrigan 2010).

Film policies are also considered as organisational innovation (Moran 1996). They became increasingly important in supporting the organisational process of film enterprise. As Wasko (2008: 55) notes "the first commission was formed in the United States during the late 1940s. This was in response to the need for film companies to have a local government liaison who could coordinate local services such as police, state troopers and highway patrols, road and highway departments, fire departments, park rangers and all of the other essential municipal and government services for shooting a production on location. As more production companies began to look beyond the limits of a regular production centre for realistic and varied locations, more cities and states began to see the need for production coordination liaison".

Finally, like other creative industries, **soft innovations**, i.e. "innovation in goods and services that primarily impacts upon aesthetic or intellectual appeal rather than functional performance" (Stoneman 2010: 22), predominantly based on intangibles, play an important role for the film industry. It is important to note, that they often depend upon or exploit other innovations, such as new product, new technology or process innovation. Hence, soft innovations are closely related to other types of film and creative innovative activity (Nesta 2009). Salt (2009) offers a detailed analysis of how technology has influenced film styles – and thus soft innovation – throughout history.

One of the most important soft innovations in the history of cinema has been film genres. As Thompson and Bordwell (2003: 228) note "many genres from the silent era continued uninterrupted into the sound period. Technical changes within the industry and social changes in the world at large, however, caused some new genres to arise and some new variants to be introduced into old ones". Many genres such as musical, screwball comedy, horror film or film noir, were invented by Hollywood. Other industries imitated the Hollywood model on a smaller scale, but as often, Hollywood's aesthetics – its dramaturgy, rules of montage and the generic happy ending – were challenged by successive waves of artistic innovation, beginning with Italian Neorealism at the end of World War II (Kapur and Wagner 2011: 81). Most historians of cinema find neorealism influential not only for its political attitudes and its world view but also for its innovations in film form (Thompson and Bordwell 2003: 362). In recent years, technology-driven soft innovations such as blue/green screen shots have also led to soft innovation by aiding the realism and believability of film narratives (Mitchell 2004: 182). Digital special effects continue to be promoted in terms of both technological and cinematic novelty, with "innovators" such as the Wachowski brothers often described as inventing a new language for cinema (Tryon 2009: 40).

In summary, the film industry has witnessed a constant flow of products, processes, organisational or soft innovations. While some of these have been attributed to particular national industry or places, less is known about the particular dynamic within particular cluster even though as mentioned they may focus on particular elements of the film value chain. This chapter adds thus to the literature on film innovation by combining a sectoral and spatial elements in looking at this dynamic.

4.4 Research Methods

Our analysis is based on data collected as part of a research on creative cluster and innovation in the UK undertaken for NESTA in 2009–2010. As part of this research we conducted an in-depth case study analysis of the innovation dynamic of the film and video production cluster in Soho, London. A preliminary analysis of the concentration of firms in the film and video sector showed that the sector was overly concentrated in London, especially in terms of production and post-production (De Propris et al. 2009). This confirmed other analysis demonstrating that the majority of firms, employment and more than 80% of the film industry turnover is located in London (BFI 2014, 2015). Within London, the neighbourhood of Soho is recognised as being the historical heart of the film industry (Pratt and Gornostaeva 2009) and still concentrated 10% of all the production firms and 14% of film distribution firms in 2007. In order to better understand the innovation dynamic of the cluster and its innovation performance, the research addressed the following questions:

- Why are film production and post-production companies located in Soho? To which extent does it support their production process?
- What is their innovation performance? How does it compare to the national average for the film industry in the UK?
- How and with whom do these firms innovate? What are their innovation inputs? Do they innovate in collaboration with other firms located in Soho? In London? Elsewhere?
- What are the critical factors supporting innovation in the cluster? How do the cultural, social and institutional contexts play a role in supporting innovation within the cluster?

To answer these questions, we adopted a mix-method and triangulation approach using both secondary and primary quantitative and qualitative data. First, in Autumn 2009, we conducted a review of secondary documentation on the film industry in Soho, London and the UK to establish the structure of the industry, its functioning, its key economic and institutional actors and the policies supporting the industry, with a particular focus on innovation policies. This secondary documentation analysis was complemented with interviews with industry experts and policy makers. Second, in early 2010, we conducted a survey with a sample of 5% of the firms in film and video production located in Soho. The survey was complemented with some follow-up interviews with a number of firms who were innovative in the previous 2 years to better understand how their innovation activities came about. In order to be able to compare the innovation performance of the cluster with the overall average of the UK industry, we use questions from the UK Community Innovation Survey [CIS] (2006–2008) and we compare our results for the cluster to the results for the industry at the national level. As such, firms were asked about their innovation activities during the period 2007–2009 as well as their overall economic activities and links with the cluster.

The CIS is part of a pan European initiative and follows the general guidelines for measuring innovation as defined in the Oslo Manual (OECD 2005). It provides information on:

- **Product innovation** – bringing to the market or into use by business, new and improved products, including both tangible goods and the provision of services. The degree of innovation is shown by the distinction between products new just to the business of which are also new to the market.
- **Process innovation**, significant changes in the way that goods or services are produced or provided, again differentiating between processes new to the business only or also new to the industry.
- **Categories of innovation inputs** include: R&D, capital goods and software acquisition, design activity, for implementing current innovations or directed to future product or process changes.
- **Organisation or management related changes**, sometimes referred to wider innovation, in strategic changes to the organisation of business or its functions, in order to achieve gains in competitiveness through efficiency or services improvements (DIUS 2008: 7).

Our analysis uses these indicators as well as an overall index indicating the use of intellectual property (IP) protection methods, i.e. "use at least one IP". In doing so, we built on Stoneman (2010)'s idea of better measuring content and/or soft innovation within the creative industries. This index indicates whether a firm has made use of any formal (registration of design, patent, trademark, copyright, confidentiality agreement) or informal (secrecy, lead-time advantage or complexity of design) intellectual property (IP) protection methods during 2007–2009. These protection methods can apply both to goods, services or processes. By encompassing all of them, we are not making any assumption about the types of intellectual property methods used by different economic sectors. Our research into the innovation within the creative industries in the UK has shown that a variety of methods are used across the creative sectors (Chapain et al. 2010). By including informal methods of protection, we also take into account the difficulty that some creative firms have in formalising their innovation outputs. Consequently, the IP index should provide an interesting overall measure of innovation outputs for the firms in the cluster.

The next section provides an overview of the British film industry in 2009–2010 and existing policies in place to support the sector at the time. This information provides the contextual information and background to the firm survey and interview findings presented in Sect. 4.5. It is important to note that some of the institutional context described in Sect. 4.5 changed after 2010, following the election of the Conservative-Liberal democrat collation. One of the major changes was the abolition of some of the regional bodies supporting the sector towards a more centralised type of support as well as the abolition of the UK Film Council created by the Labour government in 2000; most of the its functions were transferred to the British Film Institute in 2011. However, the overall way that the industry is structured and operates has broadly remained the same since then.

4.5 The British Film Industry and the Soho/London Film Cluster in 2010

4.5.1 The British Film Industry

Britain with France and the USA was one of the first countries to develop the film industry technology (Bakker 2008). As mentioned previously, today, the UK is one of the six most important producers of films in the global film market, producing more than 100 films per year, 119 in 2010 (UK Film Council 2011). The development of the film industry in Britain has been strongly influenced by the American film industry, like in many other countries across the globe (Crane 2014). In the case of the UK, the proximity in language has also made each country an important market for the other (Blair and Rainnie 2000; Baillieu and Goodchild 2002; Goldsmith and O'Regan 2005; House of Lords 2010).

Since the 1920s, the Hollywood model which has developed based on the integration between production and global distribution has allowed the USA to develop a global dominance in the distribution of films,[1] making the industry very profitable and benefiting from important financial resources (Crane 2014). Alternatively, the British film sector has developed more like a cottage industry based on small producers focusing on developing the more cultural and creative component of film production (Baillieu and Goodchild 2002). Attracted by an easier access to the British market and its talent, American film studios have invested in the UK since the beginning of the twentieth century, initiating and financing film projects but also building studio infrastructure (Goldsmith and O'Regan 2005). This is

[1]The six American studios, i.e. Warner Bros, Walt Disney, Sony Pictures, 20th Century Fox, Universal and Paramount Pictures have controlled more than 75% in the distribution market share over the period 1995–2016 see: http://www.the-numbers.com/market/distributors.

illustrated by Baillieu and Goodchild (2002: 156): "Britain undoubtedly has the creative and technical talent. London is a hotbed of expertise for post-production, and the supply of technical talent attracts the Hollywood studios to produce their blockbusters in England".

Like other countries, the UK has protected its film industry through various cultural policies; this starting in 1925. One of the key objectives of these policies has been notably to try to regulate the American film industry influence, for example by imposing quotas on the distribution of UK made films as well as using various tax incentives for UK film production; this has given rise to coproduction (House of Lords 2010). These policies have varied over the years depending on the UK government views over the importance of the UK film industry for the British economy. Combined with the evolution of US film policies and change in technology and distribution channels, notably the development of TV after the second World War, the British film industry has experienced various ups and down with various degree of dominance and presence of the US studios (Blair and Rainnie 2000; Baillieu and Goodchild 2002; House of Lords 2010). Since the end of the 1990s, the industry has experienced a renewed growth with a renewed government interest, notably by the Labour government at the end of the 1990s which created some new tax relief for British made films and new finance opportunities, notably through the UK Film Council created in 2000.

As such, the number of film and video production companies increased from 1,745 to 10,755 from 1996 to 2014 (+616%) – (see BFI 2014, 2015). This exponential increase has been supported by a growth in the UK spend on feature films which was multiplied by five from 1994 to 2010 and reached around £1.5 billion in 2014. However, like today, the great majority of this spend consisted of inward investment, predominantly from the USA, 80% in 2010; the remaining 20% was split between domestic UK features (15%) and co-production (5%).

The UK film industry is consequently operating on two different models: on the one hand, UK/US studio films which follow the traditional integrated international studio production and distribution model represented by US majors and, on the other hand, independent UK films produced by small companies involved in films production, distribution and related services. In the latter model, US studios tend to dominate the distribution side of the business as the UK is lacking international scale distributors. The UK Film Council (2009a) noted that there was only one large UK distributor (Entertainment) and the company tended to operate on the domestic market. Some companies such as Working Title or DNA have long term production-distribution arrangements with US distributors while others work on single project deal before or after production. Film exhibition in the UK is largely concentrated within five largest distributors (Odeon, Cineworld, Vue, National Amusements and Ward Anderson), which owned 74% of all UK screens in 2008 (UK Film Council 2009b). It is important to note that there are strong linkages between the film and TV industries in terms of value chain. As such, other consumption channels include DVD/video rental and sales, pay TV, terrestrial TV, free multi-channel TV and video-on-demand.

The strength of the film and video industries in the UK lies in "the depth of its production sector and skills base" (UK Film Council 2009a). Overall, film and video production and post-production accounted for 92% of all firms in the sector in 2010 and around 60% of the sector's turnover. The post-production sector has been growing rapidly since 1997 (Oxford Economic Forecasting 2005; BFI 2015) and the UK has developed and is now renowned for its strong visual effect competences (Oxford Economic Forecasting 2005). The UK also has significant number of star actors and internationally renowned directors and producers. In addition, the film and video industries benefit from some large and medium-sized studios and a plethora of small film studios nationally as well as a large service sector providing all ancillary and specialist necessary for film production and post-production (ibid.).

The film industry has received support from the UK government since the Cinematography Films Act of 1927 (UK Film Council 2009a). In 2000s, the film industry in the UK received public funding from a number of UK government, public sector and European sources; this public support amounted to a minimum[2] of £261 million in 2007/2008 (UK Film Council 2009a). As mentioned in the 2000s, the UK Film Council (now replaced by the British Film Institute) was one of the main channels for most forms of government support for film except the UK tax relief. With a budget of £70.6 million in 2007/2008 (mostly from the National Lottery and the Department for Culture Media and Sport) the UK Film Council supported activities at the various stages along the film value chain as well as other organisations supporting the sector at the national and local levels. Overall, public spend support to the film and video industries focused, in priority, on production activities (56% of total public spend), distribution and exhibition (12.4%) and training, skills and education (14%). Indeed, training and skills were an important element of the film development strategy in the UK in the 2000s as expressed in the document *A Bigger Future, the UK Film Skills Strategy* (UK Film Council and Skillset 2003). There have been a number of organisations delivering degrees (Screen Academies) or accredited training courses for the film sector across the UK.[3] Finally, the UK has benefited from a sophisticated film finance sector with specialists in the management and legal dimensions of the film economy (see UK Film Council 2009a, p. 26).

In summary, the UK film industry is well developed and characterised by the presence of the various elements necessary to the film industry value chain. It has also built strong and renowned strength in terms of production and post-production and is strongly embedded within national policies. Nevertheless, it is also tightly intertwined with the American film industry in terms of finance, production and distribution.

[2]Local government spend is difficult to capture and is not included in this figure.

[3]For more information, go to http://www.skillset.org/film/training_and_events/.

4.5.2 The Soho Cluster in 2010

In terms of geography, as mentioned earlier, Soho and London have been the heart of the film industry very early on: "The historic focus of the film industry is Soho, London – to be precise Wardour Street. It was really in the 1920s that Wardour Street was redeveloped with significant investment from US film companies seeking to secure a foothold in the UK" (Pratt and Gornostaeva 2009: 129). This historical legacy was reflected in terms of the breadth and variety of actors present in the area both industry wise and in terms of local and regional institutions in 2010.

Indeed, London cluster with Soho at its core was composed of firms and organisations covering all aspects of film development, finance, production and post-production, distribution and exhibition.[4] There also existed a large number of trade associations and representative bodies such as the UK Screen Association, Directors UK, and Association of Motion Picture Sound. UK Film Council (2009a) listed at least 27 of them in the UK, mostly located in London.

The industries benefited from at least 13 higher education institutions including Screen Academies and seven training facilities in London. The cluster was also supported by seven major national public bodies: the British Council, the British Film Institute, First Light Movies, National Film Trustee Company, Skillset, the UK Film Council and the UK Media Desk. Six out of these seven institutions were based in London. At the local and regional level, the cluster was supported by the then London Development Agency (LDA) through its general strategy to support the creative industries and through Film London, "the capital film and media agency" which offered a range of business support and training facilities through various services.[5] The LDA's creative industries strategy involved support to the London Film Festival. While the LDA had some initiatives to support innovation; these were general and not targeted especially towards the cluster. Film London supported innovation indirectly by providing some information and support in terms of:

1. Helping firms to film in London
2. Exhibiting events and facilities
3. Training and skills
4. Marketing and events
5. Production funding schemes

Nevertheless, in 2010, LDA was also running some projects such as *Own It*, an intellectual property advice service for creative enterprises and *01Zero-One*, an audio-visual training and event centre in the heart of Soho offering seminars and business advice related to the latest technology development.

Moreover, the Greater London Authority (GLA) was then also working in partnership with Film London and the LDA to support the industry as part of its

---

[4]See http://directory.filmlondon.org.uk/facilities/facixget.asp for a list of firms and organisations.

[5]See http://www.filmlondon.org.uk/.

cultural strategy. This support took the form of facilitating filming in London, financial support to the London Borough Film Fund Challenge, a scheme that gives young people training to make their first short film and thereby gain a foothold in the sector as well as other business support (GLA 2008).

In summary, in 2010, the film and video cluster in London, with Soho at its core, was diversified and strongly embedded within the local and regional institutional environment. Nevertheless, overall, there did not seem to be any comprehensive and specific strategy fostering innovation in the sector directly. Given the strong public support in terms of skills and training, production funding schemes and exhibition – we could assume that this support fostered innovation in the sector in a more indirect way.

4.6 Innovation in the Soho Cluster: Survey and Interviews' Findings

The great majority of the film production and post-production firms surveyed were small and medium sized firms with either a sole trader or private limited company status. They had been created in Soho or London; half of them since 2000s. Around 92% of them felt that they were part of a cluster. These firms display various reasons to be located in Soho (Fig. 4.1). For a fifth to a third of them, these reasons are linked to clear economic interactions along their value chain either in terms of

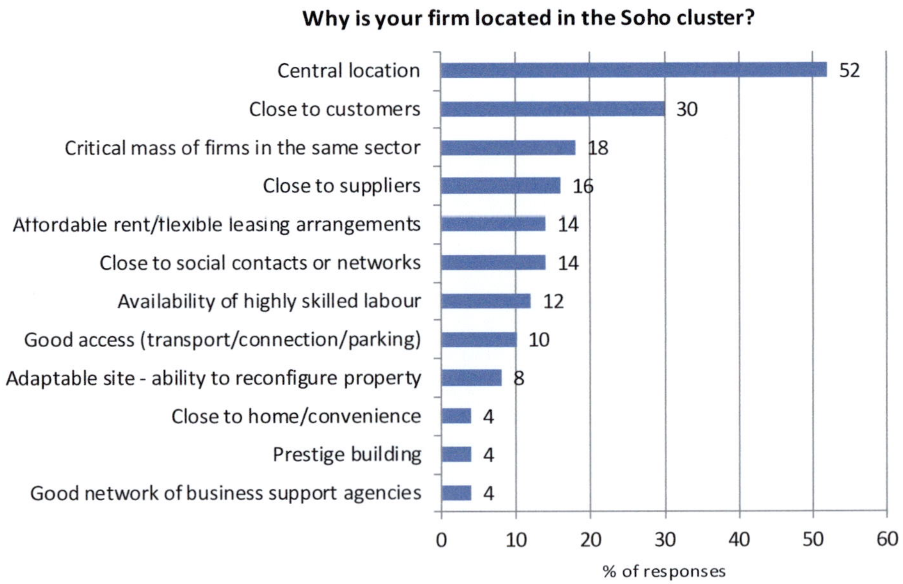

Fig. 4.1 Firms' reason to be located in the Soho cluster – 2010

suppliers (16%) or customers (30%). For some firms, their location in Soho is connected to economies of scale such as being within a critical mass of firms (18%) or wider agglomeration economies effects such as being close to their social networks (14%). However, interestingly, for most firms, the reason to be in Soho relates to its role as the central location for the film industry in London. When asked about this in interview, respondents highlighted that if you wanted to be considered as a key player of the industry you needed to be located in Soho – the location here linked to the notion of status attached to Soho as the historical heart of the industry. This is an interesting finding as other research have highlighted a process of deconcentration of some actors of the film industries towards the suburbs of London to, in reverse, regain some autonomy from the industry (Gornostaeva 2009).

Firms responses with regards to the local anchoring of their economic transactions within the cluster or within London and to the geographical origin of their workforce highlight an important degree of local embeddedness. More than a third of firm had clients or suppliers in Soho and more than 70% in London. Nevertheless, around 40% were also working with customers and suppliers located in the rest of the UK and more than 25% abroad – see Table 4.2. The latter thus demonstrates the need to also consider the importance of these external linkages in the way the cluster functions and potentially their impacts in terms of innovation. Nevertheless, comparable data on the location of the customers of UK film firms (Table 4.3) indicate that Soho firms are much more dependent on their local environment than

Table 4.2 Local embeddedness – Firms in the Soho cluster – 2010

	Location of suppliers (% of all firms)	Location of clients (% of all firms)	Recruitment of workforce (Average % all firms)
Soho	36	38	27
London	72	70	56
Rest of UK	38	40	12
Abroad	24	28	4
Total			100

Table 4.3 Location of clients, Soho cluster and UK film production sector (% of all firms)

	Soho cluster – 2007–2009 (%)	UK – motion picture and video production – 2006–2008 (%)
Soho	38	NA
London/Same UK region	70	56
Rest of UK	40	57
Abroad	28	36
Total	100	100

Source: Own data and CIS data (2006–2008)

the national average; emphasising the role of the cluster for these firms. In addition, it is interesting to note that firms in the cluster were more geographically restrained in terms of recruitment of their workforce: on average, 83% of this recruitment was made locally; this would suggest that there is a strong embeddedness of the skills necessary to the functioning of the film sector within the cluster and potentially less mobility of its inherent tacit knowledge in terms of the day-to-day firm activities.

Findings about the innovation performance and outputs of the firms located in the cluster are presented in Table 4.4 in comparison with the UK film sector as a whole. Firms in the cluster are highly innovative compared to the national average for the industry, with overall twice as many firms having introduced a product or process innovation during the 2 years prior to the survey: 58% versus 24% for the UK. In addition, more than 80% of these firms had made use of at least one intellectual property methods during that period. These innovations seem to be first about new services (44%), organisational change (40%), new processes (36%) and then new goods (20%). It is interesting to note than half of these innovations (in terms of goods, services and processes) were actually new to the market or the industry, denoting a high degree of radical innovation in the cluster. Firms in the cluster seem particularly good at introducing new innovation process, six times more frequently than the industry as a whole. In addition, the fact that half of these process innovations were also new to the industry is remarkable as usually only one third of all process innovations in the UK are also new to the industry (BIS 2010). This high performance of the Soho cluster in terms of process innovation may be

Table 4.4 Innovation outputs, Soho cluster versus UK film production sector (% of all firms)

	Soho cluster – 2007–2009 (%)	UK – motion picture and video production – 2006–2008 (%)
Introduced new or significantly improved goods	20	10
Introduced new or significantly improved services	44	21
Introduced a new good or service new to the business	38	12
Introduced a good or service new to market	22	12
Introduce any new or significantly improved process	36	6
Introduce any new or significantly improved processes new to industry	18	NA
Introduced a product or process innovation	58	24
At least one IP	82	NA
Organisation change	40	22

Source: Own data and CIS data (2006–2008)

explained by the recognised expertise of firms in the Soho cluster in visual effects and post-production. These findings would suggest a strong cluster effect in terms of the innovation performance of the Soho cluster. The question though is to what extent the innovation is taking place based on local expertise and linkages.

Table 4.5 presents the activities of the Soho firms in terms of innovation inputs in comparison to the UK industry as a whole. Overall, firms in the clusters are more prone to invest in innovation inputs compared to the national averages but the importance and patterns of investment seem to be relatively similar except for marketing activities. Four times more firms in the Soho cluster invest in marketing activities. Part of it can be explained by the linkages between the film value chain and the advertising value chain (UK Film Council 2003) but can also be linked to the intensive marketing strategy used in the film industry to increase market share (Crane 2014). As expected, technology plays an important role for the firms in the Soho cluster with almost half of them having invested in either hardware or software in the 2 years prior to the survey. This proportion is only a third for all UK firms. One of the most interesting findings reside in the importance of internal R&D for firms in the cluster – 40% of them having invested in this activity. This suggests that apart from technology and marketing, innovation seems to rely on internal resources, making the role of firm workforce even more critical. This is confirmed in Table 4.6 which shows that the main sources of innovation for the firms in the Soho cluster are: internal staff (50%), clients or customers (30%), freelancers (14%) and suppliers (12%). As firms in the cluster are highly reliant on the local labour pool and tend to trade with customers and suppliers in Soho and London, this would suggest that the knowledge and skills developed in the cluster are critical to its innovation dynamic. It is interesting to note that universities do not play a role in the innovation dynamic of the cluster. Interviews with firms suggested that given the pace of changes in technology, it can be difficult for universities to keep up to date with knowledge and potential innovation opportunities within the industry. We seem thus to be in presence, in the Soho cluster, of a privately-led dynamic of innovation. Tables 4.6 and 4.7 which present findings with regards to

Table 4.5 Innovation inputs, Soho cluster versus UK film production sector (% of all firms)

	Soho cluster – 2007–2009 (%)	UK – motion picture and video production – 2006–2008 (%)
Internal R&D	40	35
External R&D	6	8
Acquisition of external machinery	12	8
Acquisition of computer hardware	44	36
Acquisition of computer software	48	35
Acquisition of external knowledge	14	11
Training for innovation	12	17
Design	18	13
Marketing	40	11

Source: Own data and CIS data (2006–2008)

Table 4.6 Main sources of innovation, Soho cluster, 2007–2009 (% of all firms)

	Soho cluster (%)
Internal staff	50
Clients or customers	30
Freelance designers	14
Suppliers	12
Other firms (not suppliers or clients)	8
Personal contacts	8
Other firms inside group	6
Competitors	6
Trade fairs/medias	2
Universities	0
Total	100

Table 4.7 Geographical sources of innovation, Soho cluster, 2007–2009 (% of all firms)

	Soho cluster (%)
Soho	22
London	36
UK	24
EU	8
Rest of the world	20

the sources of innovation for firms confirm the importance of internal staff and the fact that when other actors are involved in the innovation process they may come from Soho or London: more than 50% of innovations originate from there. Nevertheless, links external to the cluster or to London seem to also play a role for more than half of the innovation suggesting a more complex and multi-scalar innovation dynamic – Table 4.7. This is illustrated in Table 4.8 by firm 5 whose first innovation originated from freelancers from Soho, London and the rest of the UK or firm 13 which innovated both internally and in collaboration with customers or suppliers from across Soho, London and the rest of the world. Findings from the interview confirmed both the importance of internal staff and the local milieu in the innovation process as well as access to online information and personal and professional networks operating within and outside of the cluster and the UK. Given the tight links between the British and American film industry, we would expect that some of these external linkages maybe with America or the American film value chain. Nevertheless, some of these American actors also have antennae in London and may also be part of the local milieu (Nachum and Keeble 2003a, b).

When asked about the critical success factors in promoting innovation (Table 4.9), firms in the Soho cluster highlighted four main critical areas: access to finance, as well as access to information on technology and markets and availability of qualified personnel. A fifth of firms also mentioned access to information and R&D resource in the locality. Alternatively, proximity to

Table 4.8 Examples of sources of innovation by firms, Soho cluster, 2007–2009

Firm 1	Innovation 1	Innovation 2	Innovation 3
Source	Clients or customers		
Area	Soho		
Firm 2	Innovation 1	Innovation 2	Innovation 3
Source	Clients or customers	Competitors	
Area	London	London	
Firm 4	Innovation 1	Innovation 2	Innovation 3
Source	Competitors		
Area	London		
Firm 5	Innovation 1	Innovation 2	Innovation 3
Source	Freelance designers	Internal staff	
Area	Soho, London, rest of the UK		
Firm 6	Innovation 1	Innovation 2	Innovation 3
Source	Freelance designers		
Area	London		
Firm 7	Innovation 1	Innovation 2	Innovation 3
Source	Internal staff	Cooperation with other firms, excluding suppliers or buyers	Personal contacts
Area		London	London
Firm 8	Innovation 1	Innovation 2	Innovation 3
Source	Internal staff	Freelance designers	
Area		Rest of the UK	
Firm 9	Innovation 1	Innovation 2	Innovation 3
Source	Internal staff	Clients or customers	
Area		London	
Firm 10			
Source	Internal staff	Clients or customers	Suppliers of services or goods
Area		Soho, rest of the world	Soho, London
Firm 11	Innovation 1	Innovation 2	Innovation 3
Source	Internal staff	Clients or customers	Cooperation with other firms, excluding suppliers or buyers
Area		London	London
Firm 12	Innovation 1	Innovation 2	Innovation 3
Source	Internal staff	Personal contacts	
Area		Soho	
Firm 13	Innovation 1	Innovation 2	Innovation 3
Source	Internal staff	Clients or customers	Suppliers of services or goods
Area		Soho, London, rest of the UK and rest of the world	Soho, London, rest of the UK and rest of the world
Firm 14	Innovation 1	Innovation 2	Innovation 3

(continued)

Table 4.8 (continued)

Source	Internal staff	Clients or customers	Cooperation with other firms, excluding suppliers or buyers
Area		Soho	Soho
Firm 16	Innovation 1	Innovation 2	Innovation 3
Source	Gaps in market		
Area	Rest of the UK, EU		
Firm 18	Innovation 1	Innovation 2	Innovation 3
Source	Suppliers of services or goods	Other firms inside the group	
Area	Soho, London, EU and rest of the world	EU, rest of the world	

Table 4.9 Critical success factors in promoting innovation – Soho cluster versus UK film production sector (% of all firms)

	Soho cluster – 2007–2009 (%)	UK – motion picture and video production – 2006–2008 (%)
Availability of qualified personnel	36	30
Access to finance	38	38
Access to innovation and R&D resources in the locality	20	NA
Proximity to university	2	NA
Access to information on technology	36	30
Access to information on markets	38	28
Critical mass of firms in the same industry	10	NA
Other	14	NA
Total	100	

Source: Own data and CIS data (2006–2008)

university is not considered critical. These findings emphasise again the strong privately led innovation dynamic of the Soho cluster as well as the importance of staff as vector of innovation. Nevertheless, it also suggests that the focus of British film policies in 2010, not specifically directed to film innovation per se, but focusing on training, tax incentive and finance as well as marketing and information support was appropriately playing an indirect role as enabler of an already strong innovation dynamic.

4.7 Discussion and Conclusion

This chapter has examined in detail the innovation performance and dynamic of a particular creative sector, the film sector within the British context. Our in-depth case study of Soho and our original data demonstrates that Soho is a highly innovative cluster – twice the national average – and that the cluster seems to be critical to this high performance by offering a unique creative milieu and a unique place of convergence of the key actors of the British film industry. Individuals from the cluster play a key role as carriers of the knowledge developed within the cluster but also by integrating new technological developments or in bringing their own local or international connections to innovation processes when needed. As such, they are an important component of the innovation process. Nevertheless, free-lancers, suppliers, clients as well as sources from outside Soho and London also contribute to innovation. This results either in very local or complex multi-scalar innovation dynamics from one innovation to the next. This creative milieu and innovation dynamics are also fuelled by the strong financial and production connection between the British and American film industries which seem to operate in a form of symbiosis. This symbiosis is certainly an important element defining the industry and the cluster. Nevertheless, it is based on a strong and unique British talent pool of skills and competences in production and post-production and it has not prevented the existence of an independent British film production sector operating outside of the American sphere of influence. The way the cluster and the industry function today is partly inherited from history. Nevertheless, the film industry has benefited from a strong policy support since the end of the 1990s recognising its potential and enabling its economic and innovation dynamic indirectly, notably in supporting skill development and access to finance but also in regulating the American influence allowing enough space for a British industry to develop. The institutional context has thus played an important role. As such, the Soho cluster demonstrates economic characteristics of the local buzz and global pipelines advocated by Bathelt et al. (2004) and the international linkages and institutional dominance of a hub and spoke type of district as defined by Markusen (1996). Within this hybrid model, our research raises some questions on the role of individual as carrier and integrator of knowledge both at the local and global levels in such clusters and more research should be conducted about this. In addition, while the Soho cluster demonstrates a privately-led dynamic of innovation, with little role for universities, questions can be raised with regards to the innovation dynamics and performance of other types of creative sectors and cluster for which the public sector may play a more important and direct role.

While the cluster clearly outperforms innovation in goods, services, organisation and processes, it displays a clear high performance in terms of process innovation. London has long been known for creativity in music and fashion but since the beginning of the twenty-first century it has emerged as a synonymous for specialised film technology. It is due to the boost of post-production services, in particular special visual effects (VFX), an industry that combines technological advances with artistic innovation. Post-production companies based in Soho became world-class innovators in software and solutions for visual effects. One of the catalysts for Soho's and London's success over the first decade of the twenty-first century was the so called the "Harry Potter effect". It was a result of the decision by Warner Brothers to use UK-based companies to create the VFX that helped to make the movie franchise a worldwide blockbuster. In addition to UK government support through tax credits and continuous film policy, the *Harry Potter* films helped to generate a steady stream of work for the Soho cluster, and entire British film industry. Thanks to that Soho experiences heavy concentration of top VFX companies. These companies include Framestore, Industrial Light & Magic, MPC (Moving Picture Company), Double Negative, Cinesite and Prime Focus World. Many of them hold some of the most prestigious achievements of the film industry, such as Oscar for Best Visual Effects, BAFTA Awards, or Primetime Emmy Awards.

The Soho film cluster benefits also from micro-clustering, a co-location of clusters related to media industries (film, TV, advertising or new media) as discussed by Pratt (2011). Many of these industries share parts of the value chain, especially in terms of postproduction services or special visual effects. As Pratt (2011: 132) notes "those industries that require collective labour to make them happen, ensemble pieces, have retained their need to be clustered. In fact they are tightly clustered because they are ensemble projects that are short life, therefore cannot survive unless there are a lot of similar activities in the same proximity; otherwise an individual cannot work on a semi-continuous basis. Arguably, this can be seen in film, advertising, TV and radio and new media". Strong global competition, however, challenges the advantages of this microclustering. More and more work is available in growing postproduction centres in Australia, Canada, Singapore and South Africa. However, so far, they have not yet achieved sufficient critical mass to overtake the British capital. Montreal and Vancouver posed the most serious risk because the Canadian government was offering competitive tax credits. In London's favour has been the fact that it is next to impossible to replicate its creative environment. It would be interesting thought to understand better the innovation dynamics of these emerging post production centres in comparison to London.

Acknowledgement The research presented in this chapter was supported financially by NESTA under the research conducted for the project creative cluster and innovation in 2009–2010 as well as by the Polish National Science Centre funding granted by decision no. DEC-2011/03/D/HS4/01662.

References

Acheson K, Maule CJ (2005) Understanding Hollywood's organisation and continuing success. In: Sedgwick J, Pokorny M (eds) An economic history of film. Routledge, London, pp 312–346

Aksoy A, Robins K (1992) Hollywood for the 21st century: global competition for critical mass in image markets. Camb J Econ 16(1):1–22

Baillieu B, Goodchild J (2002) The British film business. Wiley, Chichester
Bakker G (2005) Stars and stories: how films became branded products. In: Sedgwick J, Pokorny M (eds) An economic history of film. Routledge, London, pp 48–85
Bakker G (2008) Entertainment industrialised: the emergence of the international film industry, 1890–1940. Cambridge University Press, Cambridge
Bathelt H, Malmberg A, Maskell P (2004) Clusters and knowledge: local buzz, global pipelines and the process of knowledge creation. Prog Hum Geogr 28(1):31–56
Bennett B, Furstenau M, Mackenzie A (2008) Cinema and technology: cultures, theories, practices. Palgrave Macmillan, Basingstoke
BFI (2015) Film industry companies. Research and Statistics Unit, BFI, London
BIS (2010) Learning from some of Britain's successful sectors: An historical analysis of the role of government. Department for Business Innovation and Skills, London
Blair H, Rainnie A (2000) Flexible films. Media Cult Soc 22:187–204
Bloore P (2013) The screenplay business: managing creativity and script development in the film industry. Routledge, Oxon
Boix R, Hervás-Oliver JL, De Miguel-Molina B (2015) Micro-geographies of creative industries clusters in Europe: from hot spots to assemblages. Pap Reg Sci 94(4):753–772
British Film Institute [BFI] (2014) Statistical yearbook 2014. BFI, London
Chapain C, Cooke P, De Propris L, MacNeill S, Mateos-Garcia J (2010) Creative clusters and innovation: putting creativity on the map. Nesta, London
Christopherson S, Storper M (1986) The city as studio; the world as back lot: the impact of vertical disintegration on the location of the motion picture industry. Environ Plan D Soc Space 4 (3):305–320
Coe NM (2000) The view from out west: embeddedness, inter-personal relations and the development of an indigenous film industry in Vancouver. Geoforum 31(4):391–407
Coe NM (2001) A hybrid agglomeration? The development of a satellite-Marshallian industrial district in Vancouver's film industry. Urban Stud 38(10):1753–1775
Coe NM, Jones JL (2005) Beyond production clusters. Towards a critical political economy of networks in the film and television industries. In: Power D, Scott AJ (eds) Cultural industries and the production of culture. Routledge, Abingdon
Cook DA (2004) A history of narrative film. Norton, New York
Cook G, Pandit NR (2008) Clustering in the broadcasting industry. In: Karlsson C (ed) Handbook of research on cluster theory. Edward Elgar, Cheltenham, pp 274–291
Crane D (2014) Cultural globalization and the dominance of the American film industry: cultural policies, national film industries, and transnational film. Int J Cult Policy 20(4):365–382
Currah A (2003) Digital effects in the spatial economy of film: towards a research agenda. Area 35 (1):64–73
Dargis M, Scott AO (2012) Film is dead? Long live movies: how digital is changing the nature of movies The New York Times
Davis CH (2008) Screen-based media industry cluster: dimensions of performance and competitiveness. Prepared for the Ontario Ministry of Culture. School of Radio and Television Arts and Ted Rogers School of Management, Ryerson University, Toronto, Ontario
Davis CH (2011) The Toronto media cluster: between culture and commerce. In: Karlsson C, Picard RG (eds) Media clusters: spatial agglomeration and content capabilities. Edward Elgar, Cheltenham, pp 223–250
Davis CH, Creutzberg T, Arthurs D (2009) Applying an innovation cluster framework to a creative industry: the case of screen-based media in Ontario. Innov Manag Policy Pract 11(2):201–214
De Propris L, Chapain C, Cooke P, MacNeill S, Mateos-Garcia J (2009) The geography of creativity. Nesta, London
DIUS (2008) Innovation modes and productivity in the UK. Department for Innovation, Universities and Skills, London

Elsaesser T (2008) Afterword – digital cinema and the apparatus: archaeologies, epistemologies, ontologies. In: Bennett B, Furstenau M, Mackenzie A (eds) Cinema and technology: cultures, theories, practices. Palgrave Macmillan, Basingstoke, pp 226–240
Eriksson S (2011) Promotion of company and local economic growth through clusters. In: Karlsson C, Picard RG (eds) Media clusters: spatial agglomeration and content capabilities. Edward Elgar, Cheltenham, pp 30–43
Fleming Tom Creative Consultancy (2015) Cultural and creative spillovers in Europe. Report on a preliminary evidence review. http://www.artscouncil.org.uk/advice-and-guidance/browse-advice-and-guidance/cultural-and-creative-spillovers-europe. Accessed on 16 Feb 2016
Goldsmith B, O'Regan T (2005) The film studio: film production in the global economy. Rowman & Littlefield, Lanham
Gornostaeva G (2009) The wolves and lambs of the creative city: the sustainability of film and television producers in London. Geogr Rev 99(1):37–60
Greater London Authority [GLA] (2008) Cultural metropolis. The Mayor's priorities for 2009–2010. GLA, London
Hall P (2004) Creativity, culture, knowledge and the city. Built Environ 30(3):256–258
Hartley J, Potts J, Cunningham S, Flew T, Keane M, Banks J (2013) Key concepts in creative industries. Sage, London
Hellbrun J (1993) Innovation in art, innovation in technology, and the future of the high arts. J Cult Econ 17(1):89–98
House of Lords 2010 The British film and television industries. First report, Communication Committee. http://www.publications.parliament.uk/pa/ld200910/ldselect/ldcomuni/37/3702.htm. Accessed 1 June 2016
Isaacs B (2013) The orientation of future cinema: technology, aesthetics, spectacle. Bloomsbury Academic, New York
Johns J (2010) Manchester's film and television industry: project ecologies and network hierarchies. Urban Stud 47(5):1059–1077
Kapur J, Wagner KB (2011) Neoliberalism and global cinema: capital, culture, and Marxist critique. Routledge, New York
Karlsson C, Picard RG (2011) Media clusters: what makes them unique? In: Karlsson C, Picard RG (eds) Media clusters: spatial agglomeration and content capabilities. Edward Elgar, Cheltenham, pp 3–29
Kehoe K, Mateer J (2015) The impact of digital technology on the distribution value chain model of independent feature films in the UK. Int J Media Manag 17(2):93–108
Kerrigan F (2010) Film marketing. Butterworth-Heinemann, Oxford
Kindem GA (2000) The international movie industry. Southern Illinois University Press, Carbondale
Leadbeater C, Oakley K (1999) The independents: Britain's new cultural entrepreneurs. Demos, London
Lorenzen M (2007) Internationalization vs. globalization of the film industry. Ind Innov 14 (4):349–357
Margolis H, Cubitt S, King B, Jutel T (2008) Studying the event film: the Lord of the rings. Manchester University Press, Manchester
Marich R (2009) Marketing to moviegoers: a handbook of strategies and tactics. Southern Illinois University Press, Carbondale
Markusen A (1996) Sticky places in slipery space: a typology of industrial districts. Econ Geogr 72 (3):293–313
Miller T, Govil N, McMurria J, Maxwell R, Wang T (2005) Global Hollywood 2. British Film Institute, London
Mitchell AJ (2004) Visual effects for film and television. Focal Press, Oxford
Moran A (1996) Film policy: international, national and regional perspectives. Routledge, London

Mossig I (2004) The networks producing television programmes in the Cologne media cluster: new firm foundation, flexible specialization and efficient decision-making structures. Eur Plan Stud 12(2):155–171
Müller K, Rammer C, Trüby J (2009) The role of creative industries in industrial innovation. Innov Manag Policy Pract 11(2):148–168
Nachum L, Keeble D (2003a) MNE linkages and localised clusters: foreign and indigenous firms in the media cluster of Central London. J Int Manag 9(2):171–192
Nachum L, Keeble D (2003b) Neo-Marshallian clusters and global networks: the linkages of media firms in Central London. Long Range Plan 36(5):459–480
NESTA (2009) Soft innovation: towards a more complete picture of innovative change. Nesta, London
OECD (2005) Oslo manual: guidelines for collecting and interpreting innovation data. OECD, Paris
OECD (2008) Remaking the movies: digital content and the evolution of the film and video industries. OECD, Paris
Oxford Economic Forecasting (2005) The economic contribution of the UK Film post production and visual effect industry. Oxford Economic Forecasting, Oxford
Power D (2011) Priority sector report: creative and cultural industries. European Commission, Luxembourg
Pratt AC (2011) Microclustering of the media industries in London. In: Karlsson C, Picard RG (eds) Media clusters: spatial agglomeration and content capabilities. Edward Elgar, Cheltenham, pp 120–135
Pratt AC, Gornostaeva G (2009) The governance of innovation in the film and television industry: a case study of London, UK. In: Pratt AC, Jeffcutt P (eds) Creativity, innovation and the cultural economy. Routledge, London, pp 119–136
Salt B (2009) Film style and technology: history and analysis. Starword, London
Scott AJ (1998) From silicon valley to hollywood: growth and development of the multimedia industry in California. In: Braczyk H-J, Cooke P, Heidenreich M (eds) Regional innovation systems: the role of governances in a globalized world. UCL Press, London, pp 136–162
Scott AJ (1999) The cultural economy: geography and the creative field. Media Cult Soc 21 (6):807–817
Scott AJ (2000) The cultural economy of Paris. Int J Urban Reg Res 24(3):567
Scott AJ (2002) A new map of Hollywood: the production and distribution of American motion pictures. Reg Stud 36(9):957–975
Scott AJ (2004) Hollywood and the world: the geography of motion-picture distribution and marketing. Rev Int Polit Econ 11(1):33–61
Scott AJ (2005) On Hollywood: the place, the industry. Princeton University Press, Princeton
Scott AJ (2012) Finding drama in newfangled filmmaking. The New York Times
Sito T (2013) Moving innovation: a history of computer animation. The MIT Press, Cambridge
Sørensen IE (2012) Crowdsourcing and outsourcing: the impact of online funding and distribution on the documentary film industry in the UK. Media Cult Soc 34(6):726–743
Stachowiak K (2017) Geografia ekonomiczna filmu (Economic geography of the film industry). Bogucki Wydawnictwo Naukowe, Poznań
Stoneman P (2010) Soft innovation: economics, product aesthetics, and the creative industries. Oxford University Press, Oxford
Storper M, Christopherson S (1987) Flexible specialization and regional industrial agglomerations: the case of the U.S. motion picture industry. Ann Assoc Am Geogr 77(1):104–117
Storper M, Harrison B (1991) Flexibility, hierarchy and regional development: the changing structure of industrial production systems and their forms of governance in the 1990s. Res Policy 20(5):407–422
Thompson K, Bordwell D (2003) Film history: an introduction. McGraw-Hill, Boston
Tomczak P, Stachowiak K (2015) Location patterns and location factors in the cultural and creative industries. Quat Geochronol 34(2):7–27

Tryon C (2009) Reinventing cinema: movies in the age of media convergence. Rutgers University Press, New Brunswick
Tschang FT (2007) Balancing the tensions between rationalization and creativity in the video games industry. Organ Sci 18(6):989–1005
UK Film Council (2003) Post-production in the UK. UK Film Council, London
UK Film Council (2009a) Film in the UK: a briefing paper. Version 3. Research and Statistics Unit, UK Film Council, London
UK Film Council (2009b) 2009 statistics yearbook. UK Film Council, London
UK Film Council (2011) Film production in the UK. Full production report. UK Film Council Research and Statistics Unit, UK Film Council, London
UK Film Council and Skillset (2003) A bigger future. The UK film skills strategy. UK Film Council and Skillset, London
UNESCO (2016) Diversity and the film industry: an analysis of the 2014 UIS survey on feature film statistics. UNESCO, Montreal
Vang J, Chaminade C (2007) Cultural clusters, global-local linkages and spillovers: theoretical and empirical insights from an exploratory study of Toronto's film cluster. Ind Innov 14 (4):401–420
Villarejo A (2007) Film studies: the basics. Routledge, London
Wasko J (2008) Financing and producing: creating the Hollywood film commodity. In: McDonald P, Wasko J (eds) The contemporary Hollywood film industry. Blackwell, Malden, pp 43–62
Wasko J, Erickson M (2008) Cross-border cultural production: economic runaway or globalization? Cambria Press, Amherst
Zhu K (2001) Internet-based distribution of digital videos: the economic impacts of digitization on the motion picture industry. Electron Mark 11(4):273–280

Chapter 5
Understanding the Role of Professional Organisations in Supporting the Development of the Creative Industries at the Local and Global Levels: Paris as the Global Node for the Fashion Industry (1973–2011)

David Zajtmann

5.1 Introduction

The main premise of this chapter is to understand how a local/national professional organisation – i.e. the *Fédération de la couture, du prêt-à-porter des couturiers et des créateurs de mode* (Federation) – can play a determinant role in helping a city, i.e. Paris, maintains its key position within a global creative sector, i.e. the high-end fashion/Haute Couture sector.

The high-end fashion industry/Haute Couture sector is structured globally around bi-annual presentations of collections: the fashion shows. These were first introduced in the 1920s in Paris by a pre-existing professional body, the *Chambre syndicale de la couture parisienne*. The presentations of collection in Paris remain important events for the global fashion industry thanks to a long-term strategy of strong professional representation, regulation and integration of national and international key industry players carried out by the Federation over time.

Haute couture started in Paris in 1858 with the establishment of a new professional activity: the *couturier*. A *couturier* is someone who signs their creations and sells their products to a private clientele. In 1911, these professionals decided to create their own professional organisation, the *Chambre syndicale de la couture parisienne*. In the 1960, faced with new competition from newcomers in the sector both domestically with the advent of the Ready-to-wear sector and from abroad with the rise of the importance of couturiers from Milan, New York, etc. the *couturiers* decided to reshape their professional organisation by extending its remit to integrate these newcomers. This led to the creation of the Federation in 1973.

D. Zajtmann (✉)
Institut Français de la Mode, Paris, France
e-mail: dzajtmann@ifm-paris.com

© Springer International Publishing AG 2017

C. Chapain, T. Stryjakiewicz (eds.), *Creative Industries in Europe*,
DOI 10.1007/978-3-319-56497-5_5

The importance of local professional organisations and regulations in the development of creative sectors has been emphasised by Scott (2000, 2010). In his work on the interaction between creative industries and cities, he demonstrates how institutions of governance and collective action constitute a major component of the "creative filed of the city" (Scott 2010, 115). A recent special issue by Doreen and Van Heur (2014) also calls for developing a better understanding of how "cultural intermediaries" such as professional associations can help shape the development of the creative economy by influencing creative production and consumption, notably through their role of mediation. This is consistent with other research on the development of the creative industries in cities which highlight that there is a need for a "wider approach that takes into account (. . .) personal and operational connections of the creative and cultural industries within and outside the region" (Chapain and Comunian 2010, 717). This chapter contributes to this debate.

Indeed, how collections are sold is a key component in the value chain of fashion companies. For these companies, being part of a network that enables them to meet professional buyers is crucial. There is therefore a strong need for intermediaries in this industry to support points of encounters where industry players can meet. In this context, a cultural intermediary such as the *Federation* can play a key role.

Over the years, attempts to contest the dominance of Paris in the high-end fashion industry by the cities of Milan, New York and London have never really succeeded. Paris has managed to respond to this competition by offering a diversified and international environment, thanks in part to the strategies put in place by the French fashion industry's governing body, the *Federation*. The city has thus maintained its leading role in the global fashion industry. This chapter studies the strategies of the *Federation* building on the literature on meta-organisations (Arhne and Brunsson 2008).

The first section of this chapter presents an overview of the fashion industry in the World. The second section discusses some of the key literature on meta-organisation. While the third section discusses the methodology, the fourth section presents our findings. The chapter concludes by discussing the lessons and implications of our findings with regard to the sectorial and spatial dynamics of the creative industries in Europe.

5.2 The Fashion Sector in the World

5.2.1 Value Chains and Fashion Shows

Two different types of value chains have coexisted in the French fashion sector since 1973: Haute Couture established in 1868 and Ready-to-wear officially set-up in 1973 – see Tables 5.1, 5.2 and 5.3. For both segments of the industry, fashion shows play an important role. Fashion shows started in Paris in the 1920s (as shown by the creation in 1921 of the *Association de protection des industries artistiques saisonnières*, a gathering of Parisian couturiers who wanted to protect their

Table 5.1 The value chain of the Haute Couture sector (established in 1868; government regulation introduced in 1945)

1. Clients and buyers
2. "Salon" from 1868 and "fashion shows" from the 1920s
3. Parisian-based companies with a couture studio

Table 5.2 The value chain of the Ready-to-wear sector ("*prêt-à-porter*") (established in 1973)

1. Clients
2. Buyers
3. Fashion shows
4. Companies which take part in the Parisian fashion week

Table 5.3 The fashion industry segments – main distinctions

French categories	Comments
Haute Couture	Parisian specificity. Regulated by the French ministry of industry
Prêt à porter des couturiers et des créateurs de mode (Ready-to-wear)	Open to non-French brands

copyright) and became more frequent and more important from 1945 onwards after the *Haute Couture* segment became regulated by the French government.

The value chain of the Haute Couture sector can be described in the following ways (Table 5.1). Individual clients and professional buyers acquire Haute Couture products by visiting the so-called salons of the couture houses. Since the 1920s, clients have also attended the fashion shows organised by these houses. These couture houses are based in Paris and contain "ateliers" inside the city of Paris as well. In contrast, the value chain of the Ready-to-wear sector can be described as follows. Individual clients acquire Ready-to-wear products in multi-brand stores (Table 5.2). The buyers from the multi-brand stores attend the bi-annual Ready-to-wear fashion shows in Paris.

The high-end fashion sector in the world can be considered on the basis of retail characteristics. From this point of view, two main types of actors need to be considered: mono-brands stores which generally belong to large and famous brands (e.g. Christian Dior or Chanel) and multi-brand stores among which we can distinguish two categories: department stores such as Bergdorf Goodman, Selfridges, etc. (the department stores from the USA playing a major role in this category) and independent multi-brand stores such as Joyce in Hong-Kong which, while less important with regard to their turnover, are considered as *avant-garde* by professionals in the industry and, as such, have an influence in the selection of the fashion brands they decide to sell.

Key moments for the industry are when professional buyers gather for the bi-annual fashion weeks where the new collections of fashion brands are presented to professional buyers who then make their orders. The fashion weeks in Paris are the ones with the most important attendance globally. Twice a year, the main international buyers and journalists gather in Paris to attend the *prêt-à-porter* fashion shows. Being part of the Parisian fashion weeks is considered as a measure of talent recognition for international fashion brands. This has motivated Japanese brands to take part in the Parisian fashion weeks in the 1980s, same for Italian brands in the 1990s and the Russian and Asian brands in the 2000s.

Three main fashion luxury groups are dominating the industry at present: the French groups LVMH and Kering and the Swiss group Richemont. All three of them include fashion brands members of the Federation. Among the brands controlled by LVMH and which are members of the Federation are: Givenchy, Kenzo and Louis Vuitton. As for Kering, we can name: Alexander McQueen, Balenciaga and Saint Laurent Paris. Finally, as for Richemont, the brand Chloé is a member of the Federation. Hermès International also represents an important brand and is present both at the Ready-to-wear fashion week and at the menswear fashion week.

The global markets for high-end garments has become quite important and was estimated at €55–60 billion in 2012 (Delpal and Jacomet 2014). Moreover, if we add to this number, the sales for luxury goods such as leather goods, perfume and cosmetics, jewellery and watches, all luxury products associated to fashion brands, then this amount increases to €203–285 billion (Delpal and Jacomet 2014).

In France, the total turnover for the fashion industry amounts to €152 billion and represents 6% of the French manufacturing production and 550,000 jobs (APUR 2015, p. 19). In 2010, it was estimated that 37,000 of these jobs were located in the greater Paris (APUR 2015, p. 21). Chanel and Christian Dior Couture, two important international brands, are major players in the Parisian high-end Fashion industry and have been presenting collection at the Parisian fashion shows (two for *Haute Couture* and two for *Ready-to-wear*) since the 1950s.

Besides Paris, three other major cities play a significant role in the global fashion industry: Milan, London and New York. The Milan fashion weeks present Ready-to-wear brands, mainly Italian ones. However no Haute Couture (as it is proposed in Paris) presentations are made in this city and few emerging designers brands are presented within this context. The London fashion weeks are recognised as very creative, and are therefore much attended by journalists and fashion talent scouts. However, few orders are being made during these fashion weeks. The New York fashion weeks have a significant international coverage. However, the business audience at these presentations is mostly North American. As such, Paris is still playing a key global role in high-end fashion with the main buyers and journalists from the industry coming to Paris for its two bi-annual events: the *couture* weeks and the other so-called “fashion weeks”. In addition, since 1973, the Paris *prêt-à-porter* week organised by the *Federation* has also remained a key industry event for international buyers and journalists as well as emerging designers who usually present their collections at this fashion week.

5.2.2 The Role of the Professional Associations for the Sector

The development of the Haute Couture and fashion sector has been strongly intertwined with professional organisations whose roles have been to regulate the industries over time and to organise the fashion shows. As mentioned, the French professional organisations have been quite crucial in this development and the Federation created in 1973 is an important national and global player. Indeed, since 2011, the Federation has an executive committee comprised of CEOs of very important corporations or companies of the fashion industry such as Chanel, Dior or the Kering group. A lot of non-French brands have also joined the Federation in the last 40 years in order to present their collections during the *prêt-à-porter* fashion weeks. We can name among other brands: Acne Studios or Vivienne Westwood.

The Federation first has an influence associated with its regulation of the *Haute Couture* sector. To be selected in this category and to be able to present at the Haute Couture fashion shows is synonymous of standing and quality in the fashion industry, in this way, we can consider that the Federation is playing a branding role. There are three categories as regards *Haute Houture*: full membership, invited member (a temporary category with easier conditions for emerging designer brands), and finally, corresponding members for foreign brands, i.e. companies which belong to the *Fédération de la Couture*, but are not based in France (for company like Valentino, which is based in Rome, Italy).

This system makes the Federation an important intermediary for different companies: the historical actors of *Haute Couture* (mainly Chanel and Dior), the emerging designer brands and finally the foreign brands. We must add that for emerging designer brands, being part of the *Haute Couture* category enables them to sign contracts with manufacturers from the apparel industry, in Europe, but also in China.

The Federation also plays a role in the organisation of the fashion shows where the *prêt-à-porter* collections are presented. The Federation advises the CEOs of emerging designers brands and let them know when it is appropriate to present their collections. The Federation also organises or co-organises important events of the fashion industry such as the *Festival International des Arts de la Mode d'Hyères* and the *Designers Apartment*, a show room for emerging Parisian designer brands operating during the Parisian fashion weeks.

The Federation is quite specific to France. However, imitating the French set-up, a few countries have set up professional organisations specialised in high-end fashion to regulate their industry. It is in Italy that these attempts have gone the further. A *Camera Nazionale della Moda Italiana* was created in 1958. This professional body is now located in Milan and coordinates the fashion weeks taking place in Milan. It also manages an incubator. In New York, was created the Council of Fashion Designers of America (CFDA). However, this organisation did not manage the organisation of the fashion shows until 2014. In the United Kingdom, a British Fashion Council was created. This organisation coordinates the London fashion weeks. However, none of these three organisations succeeded in having a real international role. As mentioned by a study (APUR 2015, p. 19): "The London and Milan fashion weeks gather mainly British stylists for the former and Italian ones for the latter. The Parisian fashion weeks attract designers from many nationalities".

5.3 Characteristics of Meta-organisations

This chapter is dependent on the idea that meta-organisations are a specific category of organisations: organisations that gather organisations. As such, the Federation can be considered as a meta-organisation as it gathers all the key players of the fashion industry such as Chanel, Dior, Lanvin and Saint Laurent Paris. The literature on meta-organisations can help us understand its evolution and dynamic and the role it has played in supporting the fashion industry in Paris.

Research in management science into professional organisations has been enriched these last few years by the study of this particular form of organisation. The concept of meta-organisation has been developed by Ahrne and Brunsson (2008, 2010) with relevant contributions from Dumez (2008) and Gulati et al. (2012). Table 5.4 summarised some of the key points from the literature. Ahrne and Brunsson (2008) explain that meta-organisations, which are defined as organisations of organisations, have some specificities that distinguish them from classical organisations. Membership of this kind of organisation is a choice, and in parallel, exit from the organisation is completely free. Members, which are organisations themselves, maintain their own identity and are equal with respect to each other (Ahrne and Brunsson 2008, 3).

Meta-organisations have characteristics that distinguish them from organisations. First, their identity is not determined as they usually have no specific legal status (Ahrne and Brunsson 2008, 9). As a consequence, the legitimacy of these organisations is not obvious. The first consequence of organisations gathering together is the coexistence of very different members with different motivations, interests and resources. Some members may even join a meta-organisation while being opposed to it. It is usually during the starting phase that meta-organisations appear to be at their strongest. Indeed, it is easy for them, at the moment of their creation, to implement clear directives that are respected by their members. However, over time, older members are less inclined to follow rules than recent members, so this may result in a heterogeneous compliance. The existence of these meta-organisations (Dumez 2008, 4) can also be challenged by one of the important members leaving the organisation.

Second, the fact that meta-organisations are comprised of organisations and not of firms has consequences that must be taken into account. Working conditions in the meta-organisations are different from those that exist within companies. Often, meta-organisations are faced with an internal tension in that they aim to facilitate collective action by gathering together separate companies. As membership is not mandatory, the question is raised as to members' participation in the meta-organisation's activities and to their loyalty. In this view, it is self-interest that motivates

Table 5.4 Literature on meta-organisations – main characteristics

Ahrne and Brunsson (2008)	Differences in terms of membership rules and decision-making process between organisations and meta-organisations
Ahrne and Brunsson (2010)	Meta-organisations have an impact on the identity and the autonomy of their members
Gulati et al. (2012)	Legal autonomy of agents which compose a meta-organisation. A meta-organisation is characterised by the degree of openness and closeness of its frontiers (more or less high) and its degree of internal stratification (a more or less strong)

companies to become members. This creates real or potential situations of conflict between their members. This point of view aligns with Olson's perspective (1965) who relies in his analysis on the rationality of economic agents.

Third, the activity of meta-organisations is concentrated on decision-making. The act of making decision is an imperative for them. The decision-making process is made easier when the interests and values of the members are similar (Ahrne and Brunsson 2008, 107). Given the diversity of members' interests, it is easier for a meta-organisation to agree on standards rather than rules. This diversity of interests also causes meta-organisations to make decisions that are ambiguous. If they are sufficiently ambiguous, members can judge the decisions to be adapted to their own objectives, values and norms (Ahrne and Brunsson 2008, 28).

Fourth, ambiguity can also be seen in the way a meta-organisation's success is measured. It is difficult to define the criteria of success. One criteria of success may be the number of members, or the share of firms that are members in the target population. This ambiguity around the measure of success makes agreeing on change difficult and a source of potential conflict (Ahrne and Brunsson 2008, 134). Meta-organisations also have an impact on the identity and the autonomy of their members (Ahrne and Brunsson 2008, 4). In commenting on the work of Ahrne and Brunsson (2008), Dumez (2008, 3) also notes that meta-organisations have far fewer resources than their members. This makes them "structurally weak".

Finally, Gulati et al. (2012) highlight two important dimensions of meta-organisations: the first is the degree to which their frontiers are open or closed, and the second relates to their degree of inner stratification.

These characteristics and points of analysis will be used as a template to examine the role of the Federation in developing the fashion industry in Paris and maintaining the role of Paris as a global hub for fashion across the globe.

5.4 Methodology

This research is based on a longitudinal analysis of the role of the Federation in supporting the creative economy in Paris. This research studied the following period: 1973–2011. The first date was chosen because 1973 was the year the

Federation was created. The year 2011 was chosen to encompass a wide period, and also to take into account two different situations: before and after the *Haute Couture* regulations were changed in 2001.

The research questions were as follows:

- How can a city support creative workers? How can links be made with creative workers who have no previous link to the city?
- What is the role of an institution (in this case a professional organisation) in maintaining the link between creative workers and a creative city?

The research used both secondary and primary qualitative data and was based on a thematic analysis of (1) the minutes (previously unreleased) of the general assembly of the Federation and (2) ten interviews conducted between 2008 and 2012. Interviewees were chosen with the aim of capturing a diverse range of firms and also to meet people who directed firms throughout the period studied (1973–2011) and include the acting president of the *Federation* (who has been the president since 1998), its director general, as well as CEOs and former CEOs of fashion firms members of the Federation. The latter cover a diverse sample of members (past and present managers of big *couture* houses, past and present managers of design houses, past and present managers of emerging designer firms). Among the people interviewed were the former CEOs of Christian Dior Couture, Chloé International, Givenchy, Guy Laroche and Hermès International and Karl Lagerfeld, Kenzo and Yohji Yamamoto in France.

The themes of the interview questions were as follows:

- Why did you join the Federation?
- If the interviewee was the CEO of a foreign brand (not based in Paris): why did you choose Paris?

The interviews were conducted face-to-face using a semi-directive method. They lasted on average 44 min. All the interviews were recorded and transcribed into verbatim. The transcribed interviews were then coded by themes using QSRNVivo10 software based on two criteria: the frontiers of the professional organisation's member population and the frontiers of the professional organisation's activity.

The data was analysed by isolating the following themes in the verbatim:

- Reasons for joining the meta-organisation
- Reasons for staying into the meta-organisation
- Opinion about the international competition within the fashion industry

A triangulation of data was also conducted in order to better analyse the data. First, we crossed the information obtained from monographies and general assembly minutes with the information obtained from the interviews. Second, we checked the accuracy of information with several professionals for the key research themes in our study. For example, one may think that we are facing an "ancient vs modern" type quarrel between various fashion categories suggesting the competition between couturiers and Ready-to-wear as an ideological one. Our research led us to a quite different conclusion. As news magazines became a weekly occurrence, they had to find information every week rather than every month and as such tropicalise their publications in this way. Triangulation also gave interesting results concerning the reality of the threat Paris faced from Milan in the 1990s. After gathering data from three key interviews, we came to the conclusion that the threat coming from American professional buyers to stop attending Parisian fashion shows was insignificant.

5.5 Findings: How a Few Strategic Decisions from an Institution Shaped an Industry

We looked at the concept of meta-organisation in order to analyse the evolution of the *Federation* and, more broadly, the evolution of the fashion industry. In particular, we analysed this case study using the two specificities of meta-organisations highlighted by Gulati et al. (2012): the degree to which their frontiers are open or closed, and the degree of inner stratification.

The main argument is that, by making a few strategic decisions, a meta-organisation helped Paris to remain the capital city for fashion. The organisation was initially created to protect its members against counterfeiting, and to discuss work conditions with employees. Over time its role evolved significantly. The organisation became the organiser of the collections presentation weeks.

The main steps of the history of the meta-organisation are the following:

- 1973: foundation of the Federation, comprised of 3 bodies: one for *Haute Couture*, one for Ready-to-wear and one for menswear.
- 1998: official strategy from the Federation: *Haute Couture* is exclusive and national, Ready-to-wear is inclusive and international.
- 2001: softening of the regulations for *Haute Couture* by the French ministry of industry.

In the 1950s and 1960s, strong competition started coming from Italy. Initiated by Giorgini, an Italian businessman, fashion shows were organised in Florence and attracted considerable interest from American buyers. An Italian coordination and promotion body, the *Camera Nazionale della Moda Italiana* (the National Chamber for Italian Fashion), was founded in 1958. The USA also started developing its own high-end fashion industry in the 1970s. Organisations such as Calvin Klein and Ralph Lauren thrived without attending the European fashion shows.

On the domestic front, the couture industry was challenged considerably by the so-called *créateurs de mode* (emerging fashion designers), who started organising fashion shows outside the official calendar and quickly won approval from specialised and generalised magazines. Most of these newcomers came from the Ready-to-wear industry. This was the case, for example, of Kenzo, who was born in Japan in 1939 and presented his first show in Paris in 1970 (Kawamura 2004, 114–119);

Sonia Rykiel, who created her eponymous company in 1966 (Jones and Mair 2003, 402) and Agnès B., who opened her first store in Paris in 1975 (Jones and Mair 2003, 50). Although some houses did not offer Ready-to-wear (such as Balenciaga, a *couture* house which closed voluntarily in 1968), others started mixing *Haute Couture* and Ready-to-wear in their fashion shows (e.g. Cardin, Courrèges, who started his couture activity in 1965, and Yves Saint Laurent).

From the mid-1960s, *Haute Couture* professionals in Paris had to face the rise of a counter-culture of influential *créateurs* and *stylistes*. Given, among other reasons, the very high price of *Haute Couture* products for the majority of French women, consumers started to pay more attention to the less expensive products offered by these *créateurs* and *stylistes*. Fashion journalists writing for French left-wing magazines strongly supported these newcomers while totally ignoring the *Haute Couture* collections (Grumbach 2008).

The independence movement found echoes in other creative industries, where movements with the word "new" began to emerge. There emerged *nouvelle cuisine* [New Cuisine] in gastronomy (e.g. with Paul Bocuse), the *nouvelle vague* [New Wave] in cinema (among others Jean-Luc Godard) and the *nouveau roman* [New Novel] in the field of literature, with, for example, Alain Robbe-Grillet (Durand et al. 2007).

A study of the covers of the two main fashion magazines in France (i.e. all the covers of the French *Elle* and *Vogue* from 1973 to 2011) showed how Parisian *Haute Couture* lost some of its dominance in favour of *prêt-à-porter*, *stylistes* and *créateurs*. For example, in 1985 not one Parisian *couture* house could be found among the 51 covers of *Elle* magazine.

With the exception of Yves Saint Laurent, who succeeded in both *Haute Couture* and Ready-to-wear categories (and who made transgressive statements such as: "Down with the Ritz, long live the street"), *Haute Couture* houses saw their position threatened by the rise of the counter-culture of *créateurs* and *stylistes*.

All the issues mentioned above were addressed through reforms to the professional fashion institutions in France. Indeed, a strategic decision was taken in 1973: to work together with the new designers who were offering fashion outside the *Haute Couture* system. In order to do so, a new professional body was created (the *Chambre syndicale du prêt-à-porter des couturiers et des créateurs de mode)*, and the first action of the body was to organise a Ready-to-wear fashion week. Another body was also created focusing on high-end menswear, the *Chambre syndicale de la mode masculine*. Alongside, the original *Chambre syndicale de la couture parisienne,* these two new organisations constituted a newly created *Fédération de la couture, du prêt-à-porter des couturiers et des créateurs de mode* (the Federation). This new professional body included new designers who did not want to abide by the constraints of *Haute Couture* as well as foreign designers, who, from then on, were allowed to present their collections during the official calendar. As a consequence, the Parisian fashion shows became increasingly internationalised and later on, in 2001, the 1945 *Haute Couture* regulation was reformed.

This key strategic change and evolution of the structure, organisation and regulation of the professional organisations representing the fashion industry in Paris (i.e. meta-organisations), to incorporate international competitors help Paris in retaining its role as the global and key centre for fashion. Since then, key international competitors have been regularly admitted as members of the Federation. A Federation policy implemented in 1998 aimed to position the *prêt-à-porter* fashion week as both inclusive and international. Over time, two major high-end Italian brands have been admitted in the Parisian *couture* week: first Valentino and then Armani. With regard to brands from the United Kingdom, we can notice that some significant brands have chosen to be members of the Federation (among others Alexander McQueen, Paul Smith or Vivienne Westwood). Brands from emerging markets have also been regularly admitted in the Federation. This has been the case for the Indian brand Manish Arora or the brand from Taïwan Shiatzy Chen.

The French national government is also a key player of the French fashion ecosystem and has supported the Federation in various ways over the years. First, the French government has helped restricting access to the industry in establishing, since 1945, a list of authorised couturiers every year. This official governmental recognition gives the Parisian *Haute Couture* a unique position in the world. Second, until 1979, the government financed the cost of fabrics for fashion shows, as long as they were produced in France. Third, from 1981 to 1986, the government offered spaces in the *Palais du Louvre* to organise fashion shows, giving a high prestige to these events. The change in the French government in 1981, following the election of François Mitterrand, brought a new Minister of Culture (Jack Lang) in place who strongly promoted young designers, i.e. *créateurs de mode* (interview with Mr. Claude Mollard, former *délégué aux arts plastiques du ministère de la Culture*, May 21, 2012). The new Minister of Culture thought that fashion, among other French *art de vivre* industries, was one of the driving forces of France's economic power, as it had been during the reign of Louis XIV, and increased the support to the industry.

In 1982, Jack Lang, as Minister of Culture, authorised the *Chambre syndicale du prêt-à-porter des couturiers et des créateurs de mode* to make collection presentations outside one of the courtyards of the Louvre Palace (Hunter 1990). Young *cré ateurs de mode* such as Azzedine Alaïa, Jean-Paul Gaultier, Claude Montana (who presented his first collection in Paris in 1974) and Thierry Mugler were helped by the Ministry of Culture (Hunter 1990). Jack Lang also gave his personal support to young foreign fashion designers helping them to establish themselves in Paris. Other examples of governmental support include French president, François Mitterrand, hosting *couturiers*, *créateurs* and *fashion journalists* at a reception in the presidential residence on October 17, 1984. This reception left a strong impression on the foreign fashion journalists and the foreign fashion designers working in Paris, and convinced them of the prominent position of fashion in France. In 1991, the French Minister of Industry also set up a commission in order to modify the regulations for *Haute Couture*. Finally, in 2001, the French government eased the regulations of the Haute Couture category to allow some newcomers to enter this exclusive category.

To sum up, the fashion professional organisation in France was initially created to gather companies who were based and who were also producing fashion in Paris. It evolved into a federation which welcomed companies which were not based in France to help maintaining the leadership of Paris as a capital city for fashion. This has been done in agreement with the French members of the organisation and supported by the French government.

5.6 Conclusion: A Relevant Use of Symbolic Frontiers and of Stratification

This chapter examines the loss of *Haute Couture*'s pre-eminence (by "loss of haute couture's pre-eminence", we mean that until the mid-1960s, *Haute Couture* was the only global reference in regard to luxury fashion) and the arrival of new actors in high-end fashion in Paris. This transition was smoothened thanks to the Federation change of strategy and mission over time. One of the research's main findings appears to be that the Federation as a meta-organisation made a relevant use of its frontiers. As mentioned above, the professional organisation significantly changed its frontiers in 1973 by incorporating newcomers and new ways of doing high-end fashion in addition to *Haute Couture*. The way this happened could probably not have been done so easily by a state organisation because the decision to add a new fashion week in the professional body could be made very quickly and without changes for the *Haute Couture* activity.

It should also be noted that not all foreign designers who came to Paris stayed in the city permanently. However, their stay could be seen as a step towards their future careers. For example, if we consider the designers who asked to join *Cré ateurs & Industriels* (a collective fashion project initiated in 1973), it is interesting to consider their international renown and careers, such as Issey Miyake in Japan, Jean Muir (a brand founded in England in 1960) in the United Kingdom and Fernando Sanchez (an American brand created in 1973) in New York.

On the one hand, a look at fashion design in Paris in the 1970s and 1980s reveals that a lot the emerging young designers (e.g. Claude Montana, Thierry Mugler and Jean Paul Gaultier) were French. However, at the same time, a Japanese movement was flourishing in Paris with Issey Miyake, followed by (among others) Reï Kawakubo (who started presenting her collection in Paris under the Comme des Garçons label in 1981) and Yohji Yamamoto. Heavily dependent on Paris (participating in the Parisian official calendar, of course, but also contributing to artistic events such as ballets or exhibitions), these designers contributed to the internationalisation of Paris' image. In 1996, for example, Paris' fashion museum presented an exhibition called *Japonisme et Mode: 1870–1996* (Kawamura 2004). Kenzo's case was different and demonstrated what Kawamura (2004, 119) called a "complete assimilation into the French Fashion system": the companies' CEOs have all been French and the fashion label was bought by the French luxury holding LVMH in 1993.

Looking back at the theory on meta-organisation, the concepts of openness and stratification have been very helpful in understanding the role and evolution of the Federation. Openness and stratification have been matters of constant concern for the Federation. Openness was illustrated by the welcoming of emerging designers; the stratification being maintained by a strong separation between Haute Couture and *prêt-à-porter* with strong regulations maintained for the *Haute Couture* category. Besides, there was a strong difference between the number of companies in the *Haute Couture* calendar (between 10 and 15) and the *prêt-à-porter* calendar (more than a hundred). The findings fit well with Gulati et al. (2012), who highlight how the degrees of openness and stratification are two important characteristics of a meta-organisation operation.

It is also interesting to look at the use of Paris as a space by the high-end fashion industry at different levels. First, space in terms of a geographical location (with production, design studio and head offices concentrated in Paris), embodied by *Haute Couture* in general. Second, the use of specific neighbourhoods by some Parisian designers, as done, for example, by the brand Kenzo which relied on the district around *Place des Victoires* for its head office and design studio. Third, the use of Paris only as a space of exhibition during the fashion shows, as done by Italian brand such as Valentino whose studios stayed in Roma, or Acne, a Swedish brands, thriving on jeans wear products. These observations deepen our understanding of the role of specific cities in global creative sectoral and spatial dynamics but also help us understanding better the role of intermediaries such as professional organisations (or other related meta-organisations) in supporting these as recommended by Taylor (2014).

Finally, this study has implications for other creative sectors with a similar structure such as the Film industry. It demonstrates new ways of regulating a creative activity inside a country but also that international dynamics can be successfully handled by a professional organisation. Nevertheless, it is also important to recognise that similar professional organisations have been created in London, New York and Milan in order to enhance the fashion industry of their respective cities, challenging the Parisian role. We seem to be in presence of a phenomenon of mimicry. This movement raises issues for the future: will these professional organisations (and meta-organisations) become a real competitor for the French professional body?

References

Ahrne G, Brunsson N (2008) Meta-organisations. Edward Elgar Publishing, Cheltenham/ Northampton

Ahrne G, Brunsson N (2010) How much do meta-organisations affect their members? Paper presented at the SGIR 7th Pan-European international relations conference, Stockholm, September 2010

APUR – Atelier parisien d'urbanisme (2015) Nouveaux regards sur l'économie à Paris. Quelques filières d'avenir: Industries créatives/numérique/mode-design/écoactivités. APUR, Paris

Chapain C, Comunian R (2010) Enabling or inhibiting the creative economy: the role of the local and regional dimensions in England. Reg Stud 44:717–734
Delpal F, Jacomet D (2014) Economie du Luxe. Dunod, Paris
Doreen J van Heur B (2014) Taking matters into third hands: intermediaries and the organization of the creative economy, Reg Stud 1–5
Dumez H (2008) Les méta-organisations. Le libello d'Aegis 4:31–36
Durand R, Rao H et al (2007) Code and social conduct in French cuisine: impact of code changes on external evaluations. Strateg Manag J 28:455–472
Grumbach D (2008) Histoires de la Mode. Editions du Regard, Paris
Gulati R, Puranam P et al (2012) Meta-organisational design: rethinking design in inter-organisational and community contexts. Strateg Manag J 33:571–586
Hunter M (1990) Les jours les plus Lang. Odile Jacob, Paris
Jones T, Mair A (2003) Fashion new: I-D selects the world's 150 most important designers. Taschen, Köln
Kawamura Y (2004) The Japanese revolution in Paris fashion. Bloomsbury Academic, London
Olson M (1965) The logic of collective action: public goods and the theory of groups. Harvard University Press, Cambridge, MA
Scott AJ (2000) The cultural economy of cities. Sage, London
Scott AJ (2010) Cultural economy and the creative field of the city. Geografiska Annaler: Series B. Hum Geogr 92:115–130
Taylor C (2015) Between culture, policy and industry: modalities of intermediation in the creative economy. Reg Stud 49:362–373

Chapter 6
Making Your Career in Creative Industries: The Paradox Between Individual Professionalization and Dependence on Social Contexts and Professional Scenes

Bastian Lange

6.1 Introducing Entrepreneurship in the Creative Industries

This article aims at discussing the issue of entrepreneurship in creative industries. The key question is: "How do entrepreneurs in creative industries invent strategies to cope with the paradox between individual professionalization and dependence on social contexts and professional scenes, i.e. the need to be substantially connected to independent socialities, their identities, places and habits versus acting as an entrepreneur with self-interest?"

The paper refers to the striking moment when a high proportion of factual micro-entrepreneurial creative professions emerged without direct governmental support. In the last few years, the status of entrepreneurs in creative industries has been associated with a highly ambiguous situation: a "new entrepreneurship" that alludes to individualized marketing strategies, self-promotion and social hardships on the one hand, but also to skilful alternation between unemployment benefit, temporary jobs, self-employment structures and new temporary network coalitions one the other hand (McRobbie 2002; Ross 2008). Following Scott 2006 (:13), performing intense multiple and constantly shifting transaction structures in cultural-products industries means that much of the workforce becomes enmeshed in a network of mutually dependent and socially coordinated career paths. This new work ethos has been celebrated ironically with the term "digital bohème", i.e. "free agents working happily on their own" (Friebe and Lobo 2006) but has also been associated with a more critical and less-euphoric view suggesting that it also leads to precarity, social exclusion and imbalanced income (e.g. Manske 2008).

B. Lange (✉)
Georg-Simmel Centre for Metropolitan Research, Humboldt University, Berlin, Germany

University of Leipzig, Leipzig, Germany
e-mail: bastian.lange@berlin.de

© Springer International Publishing AG 2017

C. Chapain, T. Stryjakiewicz (eds.), *Creative Industries in Europe*,
DOI 10.1007/978-3-319-56497-5_6

In recent years, key urban and cultural developments such as urban creative districts, creative fairs, design weeks and creative trainings in academia, have been propelled by new cultural as well as entrepreneurial agents (Wry et al. 2011; Navis and Glynn 2010; Svejenova et al. 2007; Steyaert 2007; Neff et al. 2005; Wilson and Stokes 2005; Hjorth 2004, 2005; Sternberg and Wagner 2004).

Nevertheless, there is an ongoing debate on how context-relevant these developments are or not (Morgan and Pritchard 2004; Scott 2004; Gertler 2003; Grabher 2002a; Christopherson 2002; Strathern 1995; Knorr-Cetina 1981). In order to contribute to this discussion, I bring forwards the argument that the conceptualization of the term "scene" helps shedding light on the action logics of these cultural entrepreneurs caught in rather paradoxical circumstances, i.e. when they depend socially on a subcultural context that is principally critical towards entrepreneurial money making forms of selfish capitalization. With the term scene – increasingly used in various fields of daily routines and habits as well as in applied, social, cultural and conceptual sciences (Hesmondhalgh 2005; Blum 2001; Irwin 1977), I will demonstrate and compare the various kinds of entrepreneurial embeddedness and their structural paradoxes as expressed above in two urban settings, Berlin and Leipzig.

On the one hand, in colloquial language, the term "scene" describes and relates creative groupings and volatile collectivisations but also informal professional networks in the creative and knowledge-based industries (Hitzler et al. 2005). On the other hand, in the cultural and social sciences, concepts of the term "scene" have been developed which aim at describing the way the urban space functions (Lange 2005). Therefore, my central question will be: Can I use the concept of "scene" as an economic geography concept to understand the embeddedness of entrepreneurial scenes as an expression of the performative production of space, and as such, better comprehend the cultural entrepreneurial logic and its structural paradoxes in the creative industries?

Following this line of thinking, the plan of the contribution is as follows. Firstly, in order to adequately understand the dilemma of micro-entrepreneurs and their dependence on various forms of social contexts on the one hand, and that of individual improvement and professionalization on the other, various strands of debates are presented in Sect. 6.3. Secondly, I discuss the notion of scenes as well as four types of "paradoxes of creativity" formulated by DeFillippi et al. (2007); these paradoxes are intrinsically linked to the paradox discussed in this chapter. These scholars describe organizational dilemmas linked to epistemological problems of the study of creativity. DeFillippi, Grabher and Jones have focused on the inner-organizational dimension of emergent network-based project ecologies and their entrepreneurial and socio-spatial practices within the creative industries (DeFillippi et al. 2007; Grabher 2002a, b). The section focuses then on one of these paradoxes, the so-called paradox of professionalization. Indeed, rapidly changing project-based constellations within flexible network formations pose structural constraints especially when it comes to steps of professionalization in the professional sociology sense: As such, the fate of these creative communities of practice is shaped and partly driven by professionalization for the simple reason that they have to survive economically.

In Sect. 6.4, it becomes obvious that cultural entrepreneurs have had to invent new forms of project-based cooperation (Sydow et al. 2004; Bresnen et al. 2004) as well as specific spatial practices in order to economically, culturally and socially sustain their economic practices and livelihoods in highly risky and instable markets. This is even more relevant in harsh transformation contexts such as post-reunification Berlin and Leipzig, where very few expertise, tools and strategic guidelines existed on how and when to become an entrepreneur in a more or less undefined market such as the post-socialist period. Cultural entrepreneurial agents have thus been developing their practices in an unclear, unstructured and unstable market realm without prior knowledge and resources (Thomas 1999; White 2002).

Section 6.5 then refers to and makes use of two existing databases and recent studies on Berlin's and Leipzig's creative industries, following the attempts of the public administration to assess the economic contribution of creative industries. I will show the potential for self-organization – and thus self-governance – of creativity and creative industries in Berlin and Leipzig in respect to the concepts and issues central to this chapter. This self-organized potential of cultural entrepreneurs is linked to the activities of communities of practice that make use of Berlin's and Leipzig's specific urban fabric. As such, the "paradoxes of professionalization" that have become obvious in the case of Berlin's and Leipzig's creative industries concern, for instance, the tension between the autonomy of creative production, on the one hand, and the necessities of professionalization, on the other hand. The local entrepreneurial scenes and its communities of practices – of which most of Berlin's and Leipzig's creative industries are made of – serve both as quality evaluation circles and drivers of creativity and innovation.

6.2 Understanding Culturepreneurs and Scenes: Defining Our Key Terms

The following section introduces the concepts of cultural entrepreneurs in creative industries and their social embedding.

6.2.1 Culturepreneurs

Some scholars have invented the term "culturepreneur" which is a compound of "culture" and "entrepreneur" and was first suggested by Davies and Ford (Davies and Ford 1998 p. 13), following Pierre Bourdieu's typological notion of an entrepreneur as someone who embodies various forms of capital (Bourdieu 1986 p. 241).

"Culturepreneur" describes an urban protagonist who possesses the ability to mediate between and interpret the areas of culture and service provision. He may be characterised, first and foremost, as a creative entrepreneur, someone who runs clubs, record shops, fashion shops, galleries and other outlets, as well as someone who closes gaps in the urban landscape with new social, entrepreneurial and socio-spatial practices. Such intermediaries increasingly emerged in the gallery, art and multimedia scenes in different European metropolises, foremost in London in the 1990s (Grabher 2001). Davies/Ford (op. cit.) characterized them as the type of people who, in structural terms, are communicative providers of transfer services between the sub-systems "business-related services" and "creative scene" and, in doing so, seem to have satisfied a necessary demand.

In addition to that, creative agents, providing services for other sectors, do not only transfer and sell symbolic goods, they provide access to cultural milieus, its avant-garde-like attributes and symbolic surplus, e.g. corporate companies and other service industries like to inscribe in the internal as well as public reputation. Therefore, it is not only the symbolic product that raises the attention of various clients, but the overall symbolic surplus and its perceivable access to cultural milieus and creative scenes.

6.2.2 Scenes

So far, the term "scene" has rarely been explicitly used in cultural and social geography (e.g. O'Connor 2002), and almost not in the context of economic geographies so far, and therefore needs further explanation.

The existence of scenes and creative milieus is regarded as indispensable (Hospers 2003 p. 143; Crevoisier 2004) for the development of attractive economic sites. This can be deduced from observing that new types of media and producers of cultural goods require specific spatial qualities, which they only find in certain places (Grabher 2002a). The existence of close interactions between production areas in the urban context is nowadays seen as a crucial characteristic of a metropolis (Hesse and Lange 2007, p. 65). Entrepreneurial actors in the field of cultural or creative industries basically depend on participating in the creation of those spaces. Following Reckwitz's argument, the key factors for the change towards what is called a creative city are art scenes, which are able to enforce the city's process of self-culturalization due to their performative character (Reckwitz 2009, pp. 22–23).

This new kind of symbolic economy not only presents an affinity towards socio-cultural milieus of metropolises. It also settles in certain places of cities, as for example inner-city in-quarters, districts with a high density of gentrified and renovated old buildings providing a high living standard (such as coffee-bars offering wireless-LAN), as well as former harbour districts or fabric buildings now converted to lofts and ateliers. These places truly bring into reality the vision of space and room, especially if low rents make these places affordable, as for example in Berlin and Leipzig (Germany). Thus, these metropolises are attractive for a clientele, which mainly works under precarious conditions and therefore is not able to afford the high rents usually charged in metropolises.

6.3 Entrepreneurial Professionalization and Its Paradox Within the Creative Industries

Recently, the term "scene" has found entrance into administrative reports and consultancy studies, aiming at describing more adequately the growth of creative labour markets as well as the functionality of their adjacent networks especially in the area of creative and knowledge-based economies (Weckerle et al. 2007, p. 108ff; SMWA 2008, p. 78). In doing so, the term "scene" was applied to the behaviour of new professional actors of creative or cultural industries in regard to networks. In this context, the term "scene" describes types of exchanges established between newcomers and established actors. It refers to a certain density and a certain critical mass in communications and interactions. Apart from actors of those sub-markets, actors of established markets and administrations emphasize the increasing relevance of "creative scenes" in regard to a city's economic structure. Especially in those informal networks, processes of social and symbolic innovation can be performed, from which all branches benefit. The so-called "entrepreneurial scenes" (ibid.), interactions which provide innovation and creativity to market actors, seem to function as incubators for regional economic processes. Therefore, the presence of these kinds of entrepreneurial market actors seems to be crucial for regionally competitive strength. As both a phenomenon and a standard term, the notion of professionalization links work biographies to the formatting of sub markets and professional categories. In addition to observations concerning mobility and theoretical assumptions regarding place-binding criteria, professionalization relates creative industries directly to socio-spatial constellations, new network practices and the creation of identity. Especially rapidly changing project-based constellations within flexible network formations pose structural constraints especially when it comes to steps of professionalization. The fate of these creative communities of practice is shaped and partly driven by professionalization for the simple reason that they have to survive economically. Thus, professionalization has become a limiting context restriction that can in particular restrict creativity. In creative industries, professionalization serves, from a socio-professional point of view, three essential functions (Lange and Mieg 2008) and can be explained on three levels: a control function, an evaluation function and an expert function.

1. The inherent control function of professionalized work is currently one of the main topics of discussion in the sociology of professions (Evetts 2003; Freidson 2001). Professionalized action is generally subject to the self-control of professionals. In professional work, other common forms of organizational or institutionalized control are substituted by self-control. Professional self-control

is also at work in organizations: new forms of human resource management even assume self-control from employed professionals. Here organizational control takes on the form of "control at a distance" (Fournier 1999, p. 280) – that is internalized self-control.

2. The second function, evaluation, is closely linked to the first one. If there is today an enduring source of legitimization for professions, then it has to be based on the institutionalized control of evaluation standards for particular professional work. Classical professions (such as the medical profession or sciences) as well as new professions or professional groups (such as in the field of web design or patent auctions) attempt to define standards for professional work in their domain and to establish systems of evaluation that also include standards for professional training. Thus, professions have certain basic, socially accepted monopoles of defining work in their domains. These monopoles are variable and subject to the dynamics of changing jurisdiction in the "system of professions" (Abbott 1988).
3. The third function, the expert function of professionalized work, plays a decisive role in the domain of creative industries from two perspectives. I see not only an external expert function (towards clients and the public), but also an internal one (in the network). The internal expert function serves to differentiate and legitimate evaluation processes by identifying those professionals who set new quality standards and – equally important – who are renowned trainers or coaches in that particular professional domain. The attribution of the "experts" in the field also determines the direction of "collective" competence development of local creative economies (as professional groups). Therefore professionalization has to be considered as a process. Professionalization involves the transformation of trust regulation (from trust in single experts to trust in qualifications), the transformation of learning (from erratic individual learning to a more academy-like training) and the transformation of quality control (from individualized trust to quality reflections in globalized professional networks).

So sum this up, professionalization denotes the transformation of an occupation into a profession, that is an occupation with a certain autonomy in defining and controlling the standards of the work of its members. Furthermore, professionalization denotes the transition towards paid work that is subject to binding quality standards. In this wide sense, people and activities can be professionalized, gaining in professionalism.

In contrast to the expectation of professionalization, the high number of recently emergent creative agents in the creative industries depend to a great extent on spontaneous informal social bonds as well as network alliances that have enabled the appearance of new creative milieus. New social practices concerning the temporary organization of projects are intertwined with the production of new social places for the exchange of experiences, knowledge and expertise. To sum this up, the paradox of professionalization has become obvious in creative industries, for instance, the tension between the autonomy of creative production, on the one hand, and the necessities of professionalization on the other. The local communities of practice – of which most of Berlin's and Leipzig's creative industries are made – serve both as quality evaluation circles and drivers of creativity and innovation.

6.4 Two Contrasting Cases: Berlin and Leipzig

Continuing with the understanding of the term "scene" as presented previously, the following parts will designate parameters, which lead to a context-specific definition of entrepreneurial scenes, its structural logics and attempts to relate and to interpret their practices in respect to the introduced paradox. This presentation will be related with elements of two entrepreneurial scenes of two different contexts – Berlin and Leipzig – thus allowing for conclusions about their specific paradoxical expressions as well as first insights on how to temporarily overcome these constraints. For a better understanding of the respective situation, I start by briefly describing the context, which – as I will argue – is characterized mainly by the entrepreneurial scenes of the cultural and creative industries.

6.4.1 Methodological Prerequisites

As case-reconstruction methods provide for a wide thematic and contentual openness, they are appropriate to the present object of research (Bogner and Menz 2002a, b; Froschauer and Lueger 2002; Elliott 2002). The following analysis is based on the assumption that entrepreneurial actions in cultural and creative industries have to be embedded in social-structural as well as in social-spatial contexts. An empirical method has to provide information about the appropriation of localities as well as about the representation of the "city". The identification of the constitutional elements of Berlin and Leipzig's scenes are based on qualitative interviews and group discussions, each with 25 participants. These were selected due to their young entrepreneurial careers, i.e. acting no more than 3 years on the design market. Half of them were not originally from neither Berlin nor Leipzig, but moved to these places due to their urban appeal. By applying rules of minimal and maximal contrasting, the results were edited for a case-reconstruction (Matthiesen 2002, pp. 119–125). The interpretation focused on how market participants evaluate branch-specific contexts of interaction, how they describe the access to those materializations of informal networks and which strategies are visible concerning their positioning in a new area.

6.4.2 Context Description

Keeping in mind the question of the social-spatial condition of cultural and creative industries, the situation in Eastern Germany appears to be paradoxical: On the one hand, since 1990, de-industrialization, emigration and demographic change led to a high vacancy rate. On the other hand, punctually this opened new spaces for experiments: galleries, ateliers or temporary projects are locations, which creative folks use for their entrepreneurial actions and thus, re-introducing them to the economic cycle (Steets 2005, pp. 108–112). In those places, social networks of creative folks are manifest, claiming feasibility: in times of global economic crises, of a shrinkage in productions, of a lack of established and administrative functionaries and an increasing need of individualization, young creative folks claim new forms of collaboration and collective (economic) actions (Friebe and Ramge 2008). It is not the economic profit, which is in focus, but the implementation of individual interests in collectives, thus furthermore gaining autonomy in regard to time, sociality and symbolic winnings.

6.4.3 Entrepreneurial Scenes in Leipzig

For a long time, Leipzig was viewed as the secret cultural capital of the GDR, being a melting pot of diverse subcultures and creative actors (Farin 2002 p. 154). The image and atmosphere of the city were shaped by writers, artists and punk bands (Bismarck and Koch 2005). In Leipzig, as elsewhere, the political changes of 1989 and the globalization, which began in the 1990s, led to a reorganization of the cultural creators' networks and scenes. Informal communication networks were crucial for coping with these changes. In 2006, over 1996 small- and middle-sized companies operating in cultural and creative industries had a turnover totalling over 1.5 billion Euros. The cultural and creative industries employ about 10,500 people, which makes them a relevant factor in the labour market. Leipzig is the centre of Saxony's cultural and creative industries. Throughout Saxony, 35% of all publishing companies, 23% of all advertising companies, 36% of all film industry companies and 34% of all editorial offices are based in Leipzig (SMWA 2008, p. 17).

6.4.4 Spatial Practices of the Design and Art Scene

The micro-spatial practices of Leipzig's cultural and creative industries have to be seen in the context of a "perforated city". Leipzig's cityscape is dominated by deterioration, housing vacancy and especially inner city modernization. The city no longer corresponds to the typical European image of a dense city structure (Steets 2008, p. 167). In the following, the micro-spatial practices of the Leipzig design and art scene will be described, taking as an example the design fair Designers' Open or the Baumwollspinnerei (a former cotton mill): The Designers' Open was initiated in 2004 by the designers Jan Hartmann and Andreas Neubert from Hartensteiner Design, inspired by the Kölner Designmesse (Cologne Design Fair) and aimed at the creation of a Saxony-wide design platform. Today it serves as a central gathering point of the design industry, as a launching pad for young national and international designers, as well as a communication forum. The actors of the design scene adopt a policy that also reacts to the city's unstable spatial situation: In 2007, the main exhibition spaces of the Designers' Open were vacant former department stores and exhibition houses in inner city and central areas. In 2008, they changed their decentralizing practice of space utilization by looking for new and smaller event locations spread over the inner city, turning the whole inner city into exhibition space. This closely interrelated conception of space represents the network-based cooperation, which is essential for the scene. Furthermore, the acquisition of central areas of Leipzig represents a desire for a larger part in defining the city's image. By advertising in the city media as well as through flyers and mail addressed to the scene and beyond, the marketing measures for the Designers' Open aimed to attract a greater number of suitors, exhibitors and members of the general public so as to stimulate interest in the Leipzig design scene and increase demand for output.

In contrast to the recent and decentralizing trend within the Leipzig design scene, the art sector reacted early to the urban spatial situation, adopting a different policy with regard to the use of space and sites. Over the years, the Baumwollspinnerei, situated in Leipzig's west, became a model for the use of space, influencing other creative actors in Leipzig: low-cost purchase of former industrial areas for use as creative spaces was combined with the establishment of social networks. An essential starting point of development of the Baumwollspinnerei area was the boom of the trend-setting New Leipzig School. Today, about 80 artists work on the site, which is now professionally managed. In 2005, Leipzig's most important art galleries moved to the area, and various enterprises have settled there. (Steets 2008, pp. 174–177). So far, the working and exhibition sites in the Baumwollspinnerei have been renovated by their users only to the degree required to make work possible. The anticipated authenticity and symbolic significance of the place have thus been preserved. Members of the scene and visitors are intended to feel that the place is unfinished, and that it can and should be used in a playful and free manner. Apart from that, the unfinished aspect of the Baumwollspinnerei also indicates temporary habitation and room acquirement practices. Now these actors stopover in Leipzig while being deeply rooted in the city milieu, but keeping new ways open leading out of Leipzig.

Unlike the Leipzig design scene, the micro-spatial practices of the art scene of Leipzig and Berlin are similar: during the annual exhibitions, new locations are introduced and symbolically developed, thus attracting the attention of members and connoisseurs of the art scene. Social exclusion takes place in a subtle manner, by announcing parties and exhibitions via e-mail newsletters, or by selling works of art to specific persons revealing themselves as being part of the scene.

6.4.5 *Entrepreneurial Scenes in Berlin*

In 2010, over 25,000 small- and medium-sized enterprises belonging to the cultural and creative industries had a turnover totalling over 22 billion euros, which was more than 20% of Berlin's GDP. The cultural and creative industries employ about 180,000 people, which makes them a relevant factor in the labour market. Over 10% of the unemployed inhabitants of Berlin work in different sectors of the cultural and creative industries (Senatsverwaltung 2011). Annual fairs and frequent exhibitions emphasize these facts.

The issue of entrepreneurial practices is relevant not only to economic development and labour politics, but also to the city's image. As in other metropolitan areas, key concepts like "Capital of Talents" or "Young Creative Industries" echo one of the most dynamic potentials of urban economics. Behind concepts like "Cool Britannia" and "Generation Berlin" lie unexamined types of relations between new and initially spatially flexible occupational groups and icons on the one hand, and their temporary rooting in new local structure patterns with new spatial necessities as well as strategies of space acquisition and differentiation on the other hand.

The micro-spatial practices of the design and art scene are to be considered in the context of a city, such as Berlin, with culturally and symbolically disputed areas. The cultural dispute is about sovereignty of interpretation and space acquisition practices. Young groups of actors from expressive occupational categories have been motivated to exhibit their art in areas usually reserved for global players, due to slow growth, low investment and exaggerated planning expectations. In reclaiming urban spaces by means of temporary events (e.g. beach bars), the new cultural entrepreneurs play a key role: their resources of knowledge and experience regarding the interrelationship of networks (events) constitute spatial practices which not only provide precise distinction, but rather – by force – assume their individual understanding of the territory, and finally develop and practise another spatial theory.

This territorial practice does not only represent a playful handling of symbols, aesthetics and semiotic materials. It rather illustrates social participation and exclusion in milieus, scenes and other networks. These materials are expressive manifestations, superficial phenomena based on the claim to informal networking and continuous ascertainment of those networks. Basically, the protagonists' micro-spatial strategies point out the capability to expose immobile and apparently well-programmed spaces to temporary changes, thus reprogramming spatial functions. Offices are turned into organizational workshops and clubs, then temporarily relocated and strategically equipped with communicative attraction potentials. The necessary organizational workshop character of these enterprises shows communicative strategies to develop the micro location "office" as an interface between fluid social networks and the art scene.

The examined location – an office of the company "Greige" – points out the idiosyncratic and at the same time playful policy of attention applied by the actors. In order to make the location exist in the scene's members' heads, a specific spatial policy of renegotiation of social affiliation is adopted. Such a location can be the meeting point of an open and clearly defined group of friends, colleagues, competitors and others. Access to the location and its perception are managed by a club-like policy.

However, in this case, the common selection mechanisms of a club, a doorman refusing entry, takes on by far subtler forms. Different mediums like mouth-to-mouth propaganda, mailing lists and flyers ensure the systematic spread of information about upcoming events, inaugurations of exhibitions or new products. Apart from this information policy, a lot of effort is done to make the location sink into oblivion again, thus minimizing social affiliation.

6.5 Commonalities and Differences of Entrepreneurial Scenes in Leipzig and Berlin

In Leipzig and Berlin three relevant entrepreneurial variations and types of scenes can be identified:

1. Entrepreneurial winner scenes: First of all, Leipzig has well-known multiplicators, for instance, stable small-scale enterprises in the advertising, design and gallery sector with big influence on other scenes. They have a high transfer capability between market sub-segments. Often the drive for profit is combined with investments in non-profit activities (for instance, public art), making them an important factor for cultural sites (for instance, the design scene). In Berlin, by contrast, multiplicators like fashion label owners and gallery owners have a rather transnational orientation, giving them a big influence on the whole market. They are dependent on local networks of small-scale enterprises. These entrepreneurial elites are to a high degree locally independent. Their drive for profit is combined with investments in other areas (for instance, art), making them an important factor in Berlin as a cultural site.
2. However, both cities also show young, ephemeral micro-entrepreneurial scenes situated in rapidly changing networks of young entrepreneurs that are in a transitional stage between obtaining qualifications and finding existence-securing employment options. They are deeply rooted in city district milieus (niche structures), often coping with precarious existence levels.
3. Unlike Leipzig, Berlin has established entrepreneurial scenes on a mature level, which are more formalized than the niche scenes. Due to the size of the Berlin metropolis, they show a high degree of locally bound knowledge, production and operational capacity.

6.6 Discussion and Conclusion

The preceding explanations have shown the present and increasing use of the concept of scene in selected socio-scientific disciplines, as well as in application-oriented fields (for instance, political consulting). This chapter has concentrated on the application of the scene concept on location, space and city as well as, in a more detailed manner, on its use in cultural and creative industries. In this young segment, scenes are applied as incubators for new cultural value-adding processes, which then can influence the regional competitiveness in a positive way. The following summary focuses on the connection between scenes, space and location, as well as on location-specific articulations of scenes.

6.6.1 Process of Collectivization

In the scenes examined, expressive values like self-fulfilment, individuality, autonomy and creativity are transmitted and actualized (Koppetsch and Burkart 2002). The articulated values and connections of values have to be permanently reproduced, comparatively mobilized and frequently checked in social arenas. As a result, scenes show a demand for individualization rooted in an experience-oriented cultural practice, which generally is no fun-oriented experience at all. Moreover, the basic principle is collectivization. Entrepreneurial scenes articulate themselves in new project-based forms of cooperation with altered spatial practices and localized policies. These new practices are characterized by a playful handling of "location" and "space", where, by means of specific strategies of allocation, different scenes aggregate from time to time. Locations are thus the terrain of the post-industrial city, disputed in a symbolic way by heterogeneous groups, communities and attitudes as well as entrepreneurial scenes. By being physically present in the urban area, their members are able to integrate and position themselves. The cultural encryptions, urban legends and local narratives are often co-designed and formulated by the scenes at selected urban locations. As a result, now and in the future, scenes act as social hubs of individualized society, allowing new constellations to be tested.

Spatial practices and movement patterns intentionally stand out and keep a distance from the strategies of a centralist geography: the geographies of the ephemeral, temporary and selective presented in the examples of Leipzig and Berlin compete with the geography of symbolic-touristic and consumer-oriented centralism.

The applied practices can be viewed as tactics in the Certeauian sense, for their subtle practices of temporary resistance against the powerful economic and political structures and their disciplinary regulations can be easily illustrated. However, this political practice is not counter political in the classical sense, but rather intentionally subtle, aimed to create a (discernible) place of contradiction. The arising spatial ambivalences or breaches offer enough material for the actors to grapple with it, to become aware of and to reflect upon themselves. Hence, the scene, expressing itself at these locations, is formed by and with the charisma of this location, which is emotionally linked to the people inhabiting it.

6.6.2 Scenes and Spaces-in-Between

These conclusions have illuminated the reciprocity between the conceptual approaches to new collectivization and the concept of space. Sociologist Silke Steets has made a significant contribution for the understanding of this reciprocity in the case of Leipzig. According to her, (urban) space is an instrument for cultural providers to create temporary places with spatial micro politics (Steets 2005). These urban places are seen by Steets from the view of artists who, in times of shrinking cities, reuse and revaluate "laid-off" places which have lost their original purpose (industrial units), and turn them into clubs, pubs and galleries. This form of spatial acquirement results in ephemeral places arising from the interaction between material-physical arrangements, their narrative and visual cargo, and the actions of the users. These places find themselves in an inter-stage between private and public, legal and illegal, market and criticism (Steets 2005, pp. 115–118). The formation process of ephemeral places is based on alienation of the audience by discursive and effective blending of different types of events like workshops and cafés, and furthermore, by artists' high degree of self-organization and little structuralization of actions through fragmentary spacing, completion of space by help of various actors, and inclusion or exclusion of space by implicit readability of arrangements (Löw 2001 pp. 98–99). Yet, the latter constrains the focussed openness of these places by postulating certain know-how typical in the scene.

6.6.3 Intrinsic Logic

Both case reconstructions have shown that entrepreneurial scenes are generally based on socio-spatial differentiation practices. Yet, instead of following universal rules, they show an intrinsic logic of the city (Löw 2008). There are rules and certainties of the extraordinary of a city, of the way certain rituals, sociolects and ideologies work. Intrinsic logics can be described as the grammar of a city or place, determining what is possible in the city and where the specific within the usual is situated (Löw 2008, p. 43). This is especially true for the places and working contexts of the cultural and creative economy. Places reveal themselves through social and communicative processes. Social densification and physical presence of the scene's members is thus necessary, because the discernible relation between the actors is necessary to make Leipzig collectively discernible as a significant place. Yet, for members of the scene, these places are seen as ways to get access to the market on the basis of cultural networks. Creative actors organize exhibitions, product presentations, vernissages, finissages and parties – in other words, short-term events that co-form social space by performativity. Dominating behavioural and language codes is crucial in order to get access to these social places and events. The "'look' and 'feel' of the location" (Helbrecht 2004, p. 200) determine success. Places thus have an intrinsic logic revealing itself in local-specific and not universally homogeneous forms of collectivization.

They organize events displaying places as social fields on the urban map. These activities are aimed at obtaining social affiliation and certainty of action. They create social intensities and highly stimulating densifications which attract attention and are aimed at introducing trends, styles and codes – in short, symbolic products – into social networks. As it can be observed best in the fields of design, music, literature and fashion, these spatial policies are aimed at testing the products of cultural and creative industries – sounds, pictures, texts, techniques and graphics – for performative effects. The social and performative testing process takes place inside the temporary and transitional rooms of these creative scenes.

Berlin could offer these places, thus allowing the development of specific capabilities for the staging of a place. Leipzig began to discover these spatial options on a large scale in the last years. At these places, the performative qualities of immaterial and symbolic products can develop in social networks and can then be negotiated. Only then is it possible to generate the communicative vocabulary necessary to express the quality of a thing – the quality of a product to be tested. The case reconstructions indicate two basic, diametrically opposed processes: (1) the significance of social places attributed with intrinsic logic acting as an access to market processes and (2) the volatility of this social field of creation.

6.6.4 Professionalization

With qualitative interpretative methodologies, it was possible to show that young cultural entrepreneurs in these emerging markets attach more and more importance to informal milieu of knowledge of social networks than to codified forms of knowledge. In contrast to such codified modes of knowledge (either of a technical, management, business or organizational nature), their ability to know how, where and when to interact is of utmost importance in order to establish a professional identity as an emerging entrepreneur. But these social-network and milieu forms of knowledge have a distinct, yet non-essentialist relation to space and place. Considering creativity as a decisive source for competitive advantage, it is crucial to shift the focus to spatially relevant practices in the field of symbolic production, because symbolically designed products must be introduced, tested, adapted and upgraded first and foremost in social contexts.

The permanent creation of new genres, formats and products demanded by the market is connected with inventing narratives and social practices as well as with spatial strategies to place these products symbolically but also spatially in urban-based social places. Before being able to talk about, let alone sell and distribute products, it is necessary to invent a narration connected to the product. Apart from these opportunities to communicate, these social practices allow potential clients to be able to talk about and to experience symbolic goods in the first place. Therefore, the encounter with the immaterial good, an imperative part of the process, has to happen on the basis of affective experiences. While focussing on the way in which products are introduced into the markets of creative industries, it is important to take into consideration that immaterial products, meaning symbols, signs, sounds etc. not only have to be communicated in social networks, they should also initially be tested on the basis of their performative and atmospheric qualities. That is why products have to be placed in carefully and consciously arranged places – such as gallery openings, exhibitions, showcases, clubs, etc. – in order to allow performances to take place at all. Symbolic products therefore acquire social relevance when people experience their performative and distinctive quality first hand; they then have the vocabulary with which to talk about the quality of immaterial products, communicate it to the respective clients, write about it etc. From this perspective, producing symbols is a social process that is stimulated, fostered, orchestrated or hampered by specific organizational as well as spatial contexts.

So the category of space and spacing provides the opportunity to analyse processes of product-based symbolical upgrading as well as re-evaluation in the field of creative industries. Forming space aims at achieving a necessary degree of professional competency with which they can present their symbolic products. These spaces provide an atmospheric-based story to accompany their immaterial products. With this in mind, professional scenes need club events, galleries, exhibitions and staged office openings etc. that can be understood as temporary place-makings resulting from social formation on the urban stage. Scene-related clubbing practices not only have infiltrated the formation of professional identities as well as respective entrepreneurial strategies to access markets, they have even become a constitutive prerequisite to forming an entrepreneurial identity in the first place. Subcultural creative scenes as well as their milieu practices, which previously operated informally, have been transformed into professional scenes of design production providing a minimum of individual and collective ground for establishing the confidence needed for communicating their products and their modes of business procedures.

In respect to the theoretical reflection that substantial paradoxes (DeFillippi et al. 2007; Lange and Kalandides 2008) are crucial in determining the way sub segments of creative industries function organizationally, the empirical material provides a preliminary answer to these paradoxes. The dynamic spatial patterns I observed in the context of Berlin's and Leipzig's creative industries concern various modes of self-regulation as well as the establishment of new professional standards. The identified social networks are organized differently in that they are perpetually changing and necessarily mobile and they operate in the context of temporary working projects (Grabher 2004).

The cultural entrepreneurs can be seen as one possible answer to this growing hybridization of urban economies. Within these emerging design markets boundaries as well as socio-spatial work-relations constantly blur. This gives rise to flexible and precarious urbanites who position themselves carefully in between different systems in order to avoid getting caught in the paradoxes of long-term working models. This reflects on the one hand an administrative state body that, by and large, follows a rather standard approach towards organizing and planning labour within a given territory, while at the same time exhibits the reality of a market that is abandoning these parameters and constituting itself far beyond the administrative borders. Reacting to this paradox, cultural entrepreneurs create their own spaces of interaction where traditional borders systematically blur. Referring to Foucault's concept of heterotopia (Foucault 2000 [1984]; Hjorth 2004), one can grasp how competition and cooperation, exchange and isolation, private and public, work and leisure can temporarily co-exist and thus allow for a new understanding of how an entrepreneurial position is formed in a newly established market. They invent forms of self-organization in order to gain access to power structures based on their ability to work with socio-spatial networks and symbolic coding practices.

If scenes are not only applied in terms of time diagnostics, as a figure of the post-traditionalization of society, but also as transferable to entrepreneurial contexts of the production of symbols, it has to be recognized that the concept "scene" can be dissolved from its main context (youth and subcultures) and applied to unexpected socio-economic fields of action (e.g. creative industries). Hence, this application does not situate social determinants next to spatial ones, but allows for the so-called spatial turn in social sciences: social practices do not take place in a city, they rather embody the specifics of the city. While Ronald Hitzler conceptualizes the phenomenon "scene" as mainly non-spatial and puts social determinants next to spatial determinants, consequences arise from giving scenes, which are characterized by constructivity and performativity, a reciprocity to space. Thus, a first contextual determinant can be conceptualized. A second determinant can be found in Hitzler's point of view and in sociology of knowledge: the concept and understanding of scene is applicable on the context of entrepreneurial action and, as a result, on forms of interaction in certain markets, that is, entrepreneurial scenes in the cultural and creative industries. However, the contextual dimension arising from the spatial turn has to be conceptualized as an explanatory and differentiating dimension, thus acting as a starting point to approach contextual articulations of entrepreneurial scenes in market sub-segments of cultural and creative industries. The following aspects mark the central conclusion of this paper:

Scene has to be viewed as a social spatial, structural category. It consists of flexible, event-oriented interaction forms with high-grade internal communication characterized by implicit, shared knowledge of common temporary forms of practice.

Thereby, entrepreneurial scenes in the field of cultural and creative industries do not only provide employment options, but also structuralize entrepreneurial relation networks by spatial practices. Considering scenes exclusively in the sociological sense as staging practices, as done so by Ronald Hitzler and others, means placing social determinants next to spatial determinants, considering space as a container next to social determinants, or even completely ignoring space. Likewise, there seems to be little reason to speak of the immediateness of the urban by means of a concept of the theatrical scene, because the structuralizing effect of scene knowledge and scene experience on the social spatial level is thereby ignored. Scenes are formed around organizational and professional elites; they are no homogeneous networks.

To conclude, the potential of scene as an analytical concept of space is based on the fact that informal networks based on collocation practices can be controlled for their space-structuralizing effects.

References

Abbott A (1988) The system of professions. The University of Chicago Press, Chicago

Bismarck B, Koch A (eds) (2005) Beyond education. Kunst, Ausbildung, Arbeit und Ökonomie. Revolver – Archiv für aktuelle Kunst, Leipzig

Blum A (2001) Scenes. In: Marchessault J, Straw W (eds) Public. Special issue of public. York University, New York/London, pp 7–36

Bogner A, Menz W (2002a) Expertenwissen und Forschungspraxis: die modernisierungstheoretische und die methodische Debatte um die Experten. In Alexander Bogner, Beate Littig, Wolfgang Menz (Hrsg.), Das Experteninterview – Theorie, Methode, Anwendung. Wiesbaden, Verlag für Sozialwissenschaften. S. 7–9

Bogner A, Menz W (2002b) Das theoriegenerierende Experteninterview. Erkenntnisinteresse Wissensformen, Interaktion. In Alexander Bogner, Beate Littig, Wolfgang Menz (Hrsg.), Das Experteninterview – Theorie, Methode, Anwendung. Wiesbaden, Verlag für Sozialwissenschaften. S. 33–70

Bourdieu P (1986) The forms of capital. In: Richardson J (ed) Handbook of theory and research of the sociology of education. Greenwood Press, Westport, pp 241–258

Bresnen M, Goussevskaia A, Swan J (2004) Embedding new management knowledge in project-based organizations. Organ Stud 25(9):1535–1555

Christopherson S (2002) Project work in context, regulatory change and the new geography of media. Environ Plan A 34(11):2003–2016

Crevoisier O (2004) The innovative milieus approach: toward a territorialized understanding of the economy? Econ Geogr 80(4):367–380

Davies A, Ford S (1998) Art capital. Art Mon 1(213):12–20

DeFillippi R, Grabher G, Jones C (2007) Introduction to paradoxes of creativity: managerial and organizational challenges in the cultural economy. J Organ Behav 28(5):511–521

Elliott R (2002) Hermeneutic single-case efficacy design. Psychother Res 12(1):1–21

Evetts J (2003) The sociological analysis of professionalism. Int Sociol 18(1):395–415

Farin K (2002) Generation-kick.de. Jugendsubkulturen heute. Beck, München

Foucault M (2000 [1984]) Michel Foucault – aesthetics: essential works of Michel Foucault 1954–1864. Penguin Books, London

Fournier V (1999) The appeal to professionalism as a disciplinary mechanism. Sociol Rev 47 (2):280–307

Freidson E (2001) Professionalism: the third logic. Polity, Cambridge

Friebe H, Lobo S (2006) Wir nennen es Arbeit: die digitale Bohème oder: Intelligentes Leben jenseits der Festanstellung. Heyne, München

Friebe H, Ramge T (2008) Marke Eigenbau – der Aufstand der Massen gegen die Massenproduktion. Campus-Verlag, Frankfurt am Main

Froschauer U, Lueger M (2002) ExpertInnengespräche in der interpretativen Organisationsforschung. In Alexander Bogner, Beate Littig, Wolfgang Menz (Hrsg.), Das

Experteninterview-Theorie, Methode, Anwendung. Wiesbaden, Verlag für Sozialwissenschaften. S. 223–240
Gertler M (2003) Tacit knowledge and the economic geography of context, or the undefinable tacitness of being. J Econ Geogr 3(1):75–99
Grabher G (2001) Ecologies of creativity: the village, the group, and the heterarchic organisation of the British advertising industry. Environ Plan A 33(2):351–374
Grabher G (2002a) Cool projects boring institutions: temporary collaboration in social context. Reg Stud 36(3):205–214
Grabher G (2002b) The project ecology of advertising: tasks talents and teams. Reg Stud 36 (3):245–262
Grabher G (2004) Learning in projects, remembering in networks? Communality, sociality, and connectivity in project ecologies. Eur Urban Reg Stud 11(2):103–123
Helbrecht I (2004) Bare geographies in knowledge societies–creative cities as text and piece of art: two eyes one vision. Built Environ 30(3):194–203
Hesmondhalgh D (2005) Subcultures, scenes or tribes? None of the above. J Youth Stud 8 (1):21–40
Hesse M, Lange B (2007) Kreative Industrien – Magma und Mantra der Stadtentwicklung Das Beispiel Berlin. Kommune 2:64–69
Hitzler R, Bucher T, Niederbacher A (2005) Leben in Szenen. Formen jugendlicher Vergemeinschaftung heute, Erlebniswelten. Leske + Budrich, Opladen
Hjorth D (2004) Creating space for play-invention – concepts of space and organizational entrepreneurship. Entrep Reg Dev 16(5):413–432
Hjorth D (2005) Organizational entrepreneurship: with de Certeau on creating heterotopias (or spaces for play). J Manag Inq 14(4):386–398
Hospers G-J (2003) Creative city: breeding places in the knowledge economy. Knowl Technol Policy 16(3):143–162
Irwin J (1977) Scenes. Sage, Beverly Hills
Knorr-Cetina K (1981) The manufacture of knowledge: an essay on the constructivist and contextual nature of science. Pergamon Press, Oxford
Koppetsch C, Burkart G (2002) Passagen-Verlag Werbung und Unternehmensberatung als Treuhänder expressiver Werte? Talcott Parsons Professionssoziologie und die neuen ökonomischen Kulturvermittler. Berliner J für Soziologie 12(4):531–550
Lange B (2005) Socio-spatial strategies of culturepreneurs. The example of Berlin and its new professional scenes. Zeitschrift für Wirtschaftsgeographie (Special Issue: Ökonomie und Kultur) 49(2):81–98
Lange B, Kalandides A (2008) Governance of creative Industries in Berlin – paradoxes, places, practices. In: Wan H, Erin Y, Theresa Y (eds) When creative industries crossover with cities. Hong Kong Institute of Planners (HKIP), Hong Kong, pp 122–133
Lange B, Mieg H A. (2008) Professionalisierungswege und Konstituierungen von „Märkten" in den Creative Industries. Geogr Z 94 4: 225–242
Löw M (2001) Raumsoziologie. Suhrkamp, Frankfurt am Main
Löw M (2008) The constitution of space the structuration of spaces through the simultaneity of effect and perception. Eur J Soc Theory 11(1):25–49
Manske A (2008) Kreative Superstars. Die soziale Platzierungsstrategie der Digitalen Bohème. In: Polar (ed) Zeitschrift für Politik. Theorie. Alltag Märzausgabe. Suhrkamp, Frankfurt am Main, pp 123–126
Matthiesen U (2002) Das Feld, die Fallstudien und der "cultural turn" in der sozialwissenschaftlichen Raumforschung. In: Matthiesen U (ed) An den Rändern der deutschen Hauptstadt. Suburbanisierungsprozesse, Milieubildungen und biographische Muster in der Metropolregion Berlin – Brandenburg. Leske + Budrich, Opladen, pp 25–34
McRobbie A (2002) Clubs to companies notes on the decline of political culture in speeded up creative worlds. Cult Stud 16(4):516–531
Morgan N, Pritchard A (2004) Contextualizing destination branding. In: Morgan N, Pritchard A (eds) Destination branding. Creating the unique destination proposition. Elsevier, Oxford, pp 11–41

Navis C, Glynn MA (2010) How new market categories emerge: temporal dynamics of legitimacy, identity, and entrepreneurship in satellite radio, 1990–2005. Adm Sci Q 55:439–471

Neff G, Wissinger E, Zukin S (2005) Entrepreneurial labour among cultural producers. "Cool" jobs in "hot" industries. Soc Semiot 15(3):307–334

O'Connor J (2002) Local scenes and dangerous crossroads: punk and theories of cultural hybridity. Pop Music 21(2):225–236

Reckwitz A (2009) Die Selbstkulturalisierung der Stadt. Zur Transformation moderner Urbanität in der "creative city". Mittelweg 36. Zeitschrift des Hamburger Instituts für Sozialforschung 18 April/Mai, pp 2–34

Ross A (2008) The new geography of work: power to the precarious? Theory Cult Soc 25 (7–8):31–48

Senatsverwaltung für Wirtschaft, Technologie und Frauen (2011) Kreativwirtschaft in Berlin – Entwicklungen und Potenziale. Senatsverwaltung für Wirtschaft, Technologie und Frauen, Berlin

Scott AJ (2004) Cultural-products industries and urban economic development: prospects for growth and market contestation in global context. Urban Aff Rev 39(4):461–490

Scott AJ (2006) Entrepreneurship innovation and industrial development: geography and the creative field revisited. Small Bus Econ 26(1):1–24

SMWA (ed) (2008) 1. Kulturwirtschaftsbericht für den Freistaat Sachsen. Sächsisches Ministerium für Wirtschaft und Arbeit, Dresden

Steets S (2005) Doing Leipzig. Räumliche Mikropolitiken des Dazwischen. In: Berking H, Löw M (eds) Die Wirklichkeit der Städte. Nomos Verlagsgesellschaft, Baden-Baden, pp 107–122

Steets S (2008) Wir sind die Stadt! Kulturelle Netzwerke und die Konstitution städtischer Räume in Leipzig. Campus, Frankfurt am Main

Sternberg R, Wagner J (2004) Start-up activities, individual characteristics, and the regional milieu: lessons for entrepreneurship support policies from German micro data. Ann Reg Sci 38(2):219–240

Steyaert C (2007) Life worlds – Entrepreneuring as a conceptual attractor? A review of process theories in 20 years of entrepreneurship studies. Entrep Reg Dev 19(6):453–477

Strathern M (1995) Shifting contexts: transformations in anthropological knowledge. Routledge, London

Svejenova S, Mazza C, Planellas M (2007) Cooking up change in haute cuisine: Ferran Adrià as an institutional entrepreneur. J Organ Behav 28(5):539–556

Sydow J, Lindkvist L, Defillippi R (2004) Project-based organizations Embeddedness and repositories of knowledge: editorial. Organ Stud 25(9):1475–1490

Thomas M (1999) Neoinstitutionalismus, economic sociology und der Transformationsfall. Frankfurter Institut für Transformationsstudien, Frankfurt am Main/Oder

Weckerle C, Gerig M, Söndermann M (2007) Kreativwirtschaft Schweiz–Daten, Modelle, Szene. Birkhäuser Verlag, Basel

White HC (2002) Markets from networks: socioeconomic models of production. Princeton University Press, Princeton

Wilson N, Stokes D (2005) Managing creativity and innovation the challenge for cultural entrepreneurs. J Small Bus Entrep Dev 12(3):366–378

Wry T, Loundsbury M, Glynn MA (2011) Legitimating nascent collective identities: coordinating cultural entrepreneurship. Organ Sci 22(2):449–463

Part II
National and Regional Spatial Dynamics of the Creative Industries

Chapter 7
Multiple Endeavours Towards New Patterns of Economic Growth: The Case of Creative and Knowledge Industries in Spain

Montserrat Pareja-Eastaway and Marc Pradel i Miquel

7.1 Introduction

European countries have been facing one of the deepest economic crises in recent decades: productive structures, households, and financial systems have been struggling to overcome this critical situation for some time. The so-called creative economy, or the specialisation in sectors with high added value and with a greater capacity for job creation and economic growth, has emerged as an alternative discourse and pursuit at the European level (i.e. the Creative Industries Taskforce in the UK or the European Commission's Creative Europe programme), rather than focusing on traditional sectors of production. Spain is one of the largest European economies and it forms part of Southern Europe, where the global financial and economic crisis has strongly damaged the productive fabric and society. During the first decade of the 2000s, the country experienced high rates of GDP growth (4.2% in 2006) based on an extremely active real estate sector (Pareja-Eastaway and Turmo 2013). Since the pillars for growth in Spain were not sustainable, the negative shock created a new scenario of considerable unemployment (26.3% in 2013, the highest peak during the period) and negative GDP growth rates (−3.6% in 2009).

The prominence at the country level of economic rhetoric based on creativity, knowledge, or innovation is of paramount importance for cities and regions as these sectors tend to be especially connected to urban environments (Flew 2010; Bontje

M. Pareja-Eastaway (✉)
GRC Creativity, Innovation, and Urban Transformation, Department of Economics, Faculty of Economics and Business, University of Barcelona, Barcelona, Spain
e-mail: mpareja@ub.edu

M. Pradel i Miquel
GRC Creativity, Innovation, and Urban Transformation, Department of Sociology, Faculty of Economics and Business, University of Barcelona, Barcelona, Spain
e-mail: marc.pradel@ub.edu

© Springer International Publishing AG 2017

C. Chapain, T. Stryjakiewicz (eds.), *Creative Industries in Europe*,
DOI 10.1007/978-3-319-56497-5_7

et al. 2011; Pratt and Hutton 2013). The consequences of the economic downturn have a visible face in the local arena: unemployment, the closing of companies and businesses, and the rise of local public debt have been major indicators of the decrease in local welfare. European cities and regions have reacted in dissimilar ways to the negative impact of the economic crisis (Musterd and Kóvacs 2013). In Spain, there is no clear city model as the country's economic diversity is also found in the configuration of its cities.

The literature offers several perspectives on the relevance of creative and knowledge sectors in the new economy since the crisis (Pratt and Hutton 2013; Pratt 2012; Flew 2010). Different approaches and disciplines provide a comprehensive analysis of the contribution of creativity and knowledge to cities and regions. In this chapter, we propose a three-axis model analysis of creative and knowledge sectors in the local arena: we consider actors, pathways, and policies as representatives of the role played by context in the development of these sectors. The aim of this chapter is to identify new patterns and dynamics of economic specialisation and development in Spanish cities and regions based on creativity and knowledge. To do this, we will analyse such patterns in four Spanish cities taking into consideration their historical development, the constellation of actors, and their governance framework. The four cities selected for analysis are the most relevant in terms of population and economic performance: Barcelona, Bilbao, Madrid, and Valencia. Madrid and Barcelona are the two largest metropolitan regions in Spain both in terms of population and GDP contribution, followed by Valencia. Despite Bilbao being the sixth largest metropolitan area after Malaga and Seville in terms of population, it is an important economic centre. In Madrid and Barcelona, the dynamics of concentration, density, and exchanges are greater/more important, which also explains their more considerable contribution to GDP. Because of this, we will mainly focus on these two urban agglomerations, illustrating the comparison with reference to Bilbao and Valencia. The uneven development of creative and knowledge industries as well as different trajectories and initiatives at the local level will be noted.

The chapter will first present a theoretical section on the state of the art concerning the dynamics of local environments with regard to the creative and knowledge economy. We will focus on the three dimensions used to examine the particular examples in the four Spanish city-regions. Secondly, the chapter assesses the relevance of creative and knowledge sectors in Spanish cities through the analysis of quantitative data, showing the uneven development of such industries in the four cities. Following this, the article shows the different institutional contexts of these cities, focusing on pathways and actors. Finally, we analyse the emergence of new public policies and initiatives to foster the creative economy emerging in these cities.

7.2 A Three-Axis Model Approach to the Development of the Creative Industries in Cities

Different approaches have framed the debates on the role of the creative economy in cities. Departing from analyses of innovation, some authors, mainly focused on economic geography, study the role of cities as spaces where creativity and innovation take place and influence economic activity (Florida 2002; Scott 2008, 1997; Storper 2013). These authors explain new geographic patterns of economic growth focusing on the role of innovation and creativity in cities and how they concentrate unevenly in different conurbations. This approach, largely based on US data, has brought about debates on whether creativity is an individual asset or the result of a social process, as well as debates on the logics of economic concentration: are companies settling where the creative class is concentrated or are workers following companies (Storper 2013)? All these debates have brought increasing attention to creativity from policy-makers in many countries, who during the last decade have been promoting policies to foster creativity in cities. In most of the cases, governments at different levels have developed lists of 'creative sectors' that have been targeted with strategic development policies to strengthen them. Nevertheless, these attempts have collided with previous understandings of sectors, or have overlapped with previous definitions. For instance, the inclusion of the fashion industry within the definition of the creative sectors while part of this industry might still be considered traditional.

During the Spanish economic boom (1997–2007), promoting the creative and knowledge economy was not central for the national government, and there were no comparable initiatives to those taking place in the UK. Literature and policy initiatives on the creative city had a certain impact in local and regional governments, although the development of such initiatives was strongly conditioned by an economic growth model based on the speculative real estate boom. Mirroring approaches in other countries, local and regional governments implemented programmes to foster classic cultural industries (cinema and television, publishing, performing arts, etc.) and some specific sectors (design, video games, etc.) under the rubric of creative city promotion. These approaches roughly followed the definition of creative industries used by the UK government's Department of Culture, Media and Sports since 1998.[1] Following this definition, the sectors included in the creative economy are: publishing, architecture, arts and antiques trade, designer fashion, video, film, music and photography, music and the performing arts, advertising, computer games, software and electronic publishing, and radio and television (DCMS 1998; Musterd et al. 2007). With the crisis and the

[1]The UK's definition of the creative industries ('those industries that are based on individual creativity, skill and talent with the potential to create wealth and jobs through developing intellectual property') includes 13 sectors: advertising, architecture, the art and antiques market, crafts, design, designer fashion, film, interactive leisure software (i.e. video games), music, the performing arts, publishing, software, and television and radio.

imposition of austerity policies in Spain, such sectors became a possible basis for new economic growth and, as we will see later, some figures certainly reveal the relevance of knowledge-intensive sectors in cities' economic performance. However, since cultural sectors were strongly dependent on public investment, efforts were directed at promoting knowledge-intensive sectors rather than the creative economy as such, although differences are found between cities in this regard. Nevertheless, the implementation of comprehensive policies to foster these sectors was limited.

In our research, we have focused on the uneven development of the creative economy in Spanish cities, analysing not only data on creative and knowledge sectors but also their institutional arrangements and the impact of such economic activities on city governance. To do this, we depart from an institutionalist perspective, focusing on actors, rules, norms, and behaviours that are historically consolidated and that explain current forms of organisation and political choices. Research has shown that the development of creative and knowledge sectors depends largely on the historical development of the city; its own pathways (Musterd et al. 2007). All cities trying to foster creative industries have developed narratives about the historical development of creativity in the city, often going back to ancient times of craftsmanship and guilds during the Middle Ages or the first years of the Industrial Revolution. The construction of such narratives runs parallel to the demand for more autonomy that cities have developed since the second half of the 1990s, strongly based on the myth of the European city of the Middle Ages: city-states in which freedom and creativity emerged in a context of tyranny (Le Galès 2002). Nevertheless, the relation between historical development and the possibilities for the establishment of creative industries is much more complex. Pathways of industrialisation and the historical role of the city in national and regional economic geography are key to understanding how the economic shift towards creativity and knowledge takes place. In addition, the presence of relevant economic actors trying to maintain the importance of certain economic activities and the existence of informal rules or forms of governing must be stressed. In this regard, the historically developed institutional context reveals the existence of norms, values, and traditions that frame the actions of local actors.

This institutional context is not static but is continuously transforming. Actors have the autonomy to change and transform it through their actions and new forms of organisation. Research has shown how actors in the creative industries develop new forms of organisation and work that mean new forms of production and new relationships with local administrations and other private actors, resulting in an impact on local government and, consequently, the transformation of the institutional context. Creative sectors might influence local policies and the development of the city like any other sector, but their forms of organisation differ: creative activities bring about the emergence of 'communities of practice' based on sharing ideas and knowledge, but also professional advice and political practices (Lange et al. 2010, p. 310). Networks and their institutionalisation shape professionalisation patterns with processes that diverge from the traditional organisation of companies and economic interests. In this regard, some authors have focused on the 'ethos' of creative production, stressing the relevance of craftsmanship in certain activities, where values linked to 'doing things well' are more important than efforts to maximise benefits (Sennett 2008). For that reason, these activities are based on the organisation of work around workshops, 'co-working spaces', or 'makerspaces' rather than traditional offices or factories, in a pattern more similar to the organisation of craftspeople.

All of this has an impact on local governance. Actors from the creative economy have different forms or organisation and professionalisation, giving rise to a different configuration of collective actors that influence policy-making. In comparison to knowledge-intensive industries, the creative industries tend to use informal networks and less institutionalised actors, which can leave them with a more limited capacity to interact with local governments. At the same time, these networks are more flexible and can develop bottom-up initiatives that connect with other citizens' demands (see for instance Pradel i Miquel 2016; D'Ovidio and Pradel i Miquel 2012; Novy and Colomb 2012; Martí-Costa and Pradel i Miquel 2012). The role and the organisation of actors involved in creative activities strongly depend on the existing constellation of actors in the city and the influence capacity that these economic activities may have. One of the main features during the economic boom in Spain was the strong role that the real estate and finance sectors had in the economy, leaving less room for attention to other economic activities. Moreover, the aims and scope of the city in the framework of territorial competition is also key. In Spain, Madrid and Barcelona tried to compete in the international arena, whereas other cities tried to remain relevant at the national or European level.

All of these elements can be operationalised considering different dimensions that are useful for understanding the consolidation of creative and knowledge sectors as cornerstones for economic growth in cities. In our empirical research, we will use the following three dimensions:

(a) The historical **pathways**
(b) The constellation of economic, political, and social **actors** in the city and the organisation of their interests in creative and knowledge activities
(c) The **policies** for creative and knowledge industries emerging from the local governance framework

In the next section, we will provide some contextual information related to the creative and knowledge sectors in the four metropolitan areas. We will pay special attention to the evolution of these sectors during the period of the economic crisis (2008–2014). As we will see, there are significant differences between cities and sectors. In Sect 7.4, we will discuss potential explanations for this divergent development according to the dimensions mentioned above.

7.3 The Emergence and Weight of Creative and Knowledge Sectors in Four Spanish Cities

After the devastating effects of the crisis, the need to reinvent the Spanish economic model was acute. Until this time, the role of creative and knowledge industries as cornerstones for a new wave of economic growth in Spanish cities received scarce attention from policy-makers and researchers in comparison to other countries. In fact, so far, there is no official definition in Spain of which sectors are involved in the creative industries. The former specialisation in a single sector had proved highly dangerous in periods of general economic downturn; therefore, the diversification of the productive fabric became a target for many regional and local agendas. Among other strategies, the reconsideration of theories based on endogenous development (Camagni 2005) took place in Spain, placing value on the unique characteristics of the territory and its economic activity. Thus, strategies to create new advantages or to build territorial specificities became increasingly important. The so-called 'knowledge economy', with a relevant percentage of highly qualified jobs, emerged as a common target for many local agendas as it was considered that this type of specialisation might reduce urban vulnerability and decrease unemployment. Several researches have attempted to explore the scope and potentialities of these sectors as major drivers of an alternative to the traditional productive model in the Spanish economy (Boix et al. 2010; Recio 2010, Sánchez-Moral et al. 2014; Méndez and Sánchez-Moral 2011; Méndez et al. 2015).

According to Sánchez-Moral et al. (2014), creative industries in Spain represent 5.18% of total employment and are concentrated in large urban agglomerations. With regard to these industries, the authors highlighted a controversy in terms of the distinction between creative industries and creative occupations, considering the creativity associated with each worker regardless of the classification of the industry they work in. In fact, if the focus is on the worker's occupation rather than the economic activity, the creative economy is more significant: for each job in the creative industries proper, there are approximately three professionals in creative occupations (Sánchez-Moral et al. 2014). During the first phase of the economic crisis (2007–2010), knowledge industries were able to generate net jobs (+5.14%) in a general context of job destruction (−10.90% in other activities) (Méndez et al. 2015). However, in the subsequent phase (from 2010 on), knowledge-intensive industries reversed that trend and joined the decline although more moderately (−1.24% vs. −9.55% for the rest). These activities had evolved in a countercyclical way, but the consequences of Spanish austerity measures driven by the European Union to ease investor fears about Spanish debt affected them very negatively. The crisis partially revealed a phenomenon that was long-standing in the Spanish economy: between 1996 and 2007, Spanish GDP increased above the European average, producing new jobs but not increasing productivity. This fact indicated a certain deficiency in the Spanish productive system in terms of achieving efficient use of the factors of production, particularly human capital. The existence of over-education, understood as the percentage of people with tertiary education or above who hold semi-skilled or low-skilled jobs, highlights the mismatch between the training of the workforce in Spain and the production system, which is unable to create jobs to absorb qualified human capital (Pareja-Eastaway and Turmo 2013; García-Montalvo and Peiró 2008).

Besides, a generalised concern was the dichotomy established between traditional sectors such as those related to real estate development (i.e. furniture, construction, appliances, etc.) and emerging ones, for instance, those based on creativity and knowledge. According to Pareja-Eastaway and Turmo (2013), ICT, financial intermediation, and telecommunications were the sectors that, between 1996 and 2007, exhibited the highest percentages of annual growth in terms of added value (9.4, 6.5, and 6.4, respectively). For construction it was 5.17%. The increase of connections between new activities and traditional sectors could be a success factor: urban agglomerations that have facilitated synergies between traditional manufacturing sectors and emerging creative and knowledge sectors are better positioned (for instance, Bilbao) than others.

Creative and knowledge sectors[2] in Spain are unevenly distributed across the territory (Méndez et al. 2012; Lazzaretti et al. 2008). Our focus will be on the four major metropolitan agglomerations in Spain, considering their similarities and differences together with their successes and failures. We will use data at the Autonomous Community level provided by the National Statistics Institute's (INE) Central Companies Directory, considering that each urban agglomeration studied acts as a driver for the overall region it is located in. These four city-regions are currently dealing with the devastating consequences of the economic crisis alongside the local challenges of a new global order. According to our definition, in the period 2008–2014, on average companies belonging to the creative economy represented around 4.5% in Catalonia, 3% in the Basque country, 3.5% in Madrid, and 4% in Valencia (see Fig. 7.1).[3] Knowledge-intensive companies show a higher percentage in all cases (around 20%), with the exception of Madrid, where they account for 25%.

Creative and knowledge companies provide employment in a dissimilar way (see Fig. 7.2). While employment in creative companies represents around 5% in Catalonia, 2% in the Basque Country, and 5% in Madrid and Valencia, knowledge-intensive companies produce a higher number of jobs in all cases: 28% in Catalonia, 22.5% in the Basque Country, 30% in Madrid, and 16% in Valencia. Throughout the period 2008–2014, creative employment over the total diminished in the four territories, but especially in those where it had higher participation over the total:

[2]Defining creative and knowledge sectors: creative industries – publishing, architecture, arts/antiques trade, designer fashion, video, film, music and photography, music and the visual and performing arts, advertising, computer games, software and electronic publishing, radio and television; knowledge-based industries – information and communication technology, finance, law and other business services, R&D and higher education.

[3]Due to methodological problems (lack of available data for city-regions), data used in this section refer to the Autonomous Communities (A.C.) (Catalonia, the Basque Country, Madrid, and Valencia), our four case studies being their capitals.

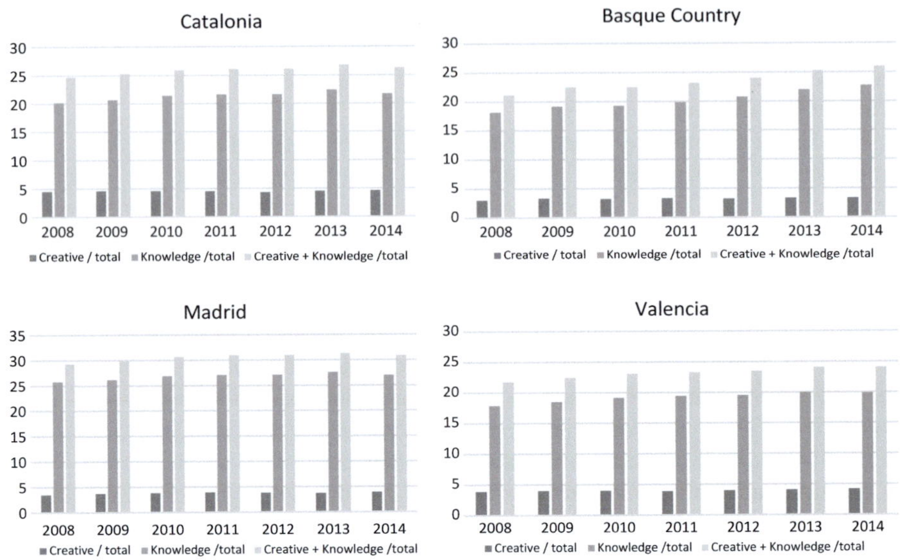

Fig. 7.1 Creative and knowledge companies in Catalonia, the Basque Country, Madrid, and Valencia (Source: National Statistics Institute's (INE) Central Companies Directory)

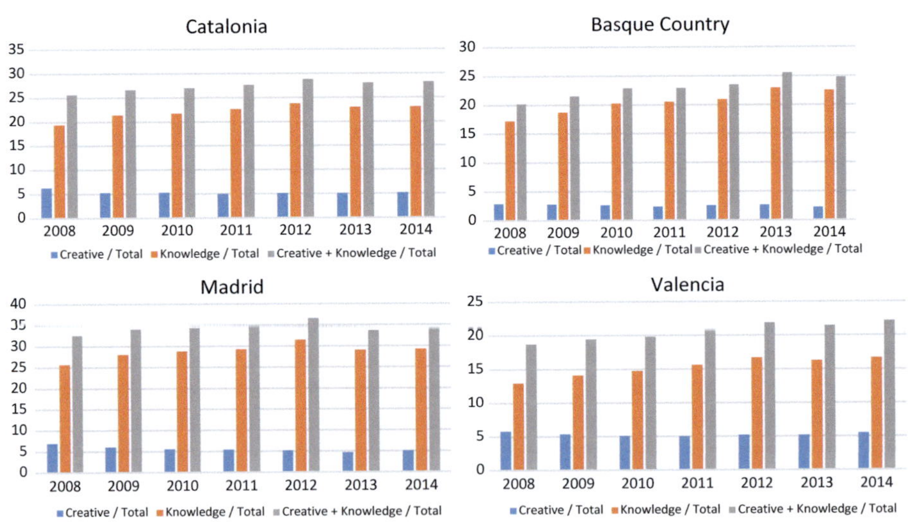

Fig. 7.2 Creative and knowledge employment in Catalonia, the Basque Country, Madrid, and Valencia (Source: National Statistics Institute's (INE) Central Companies Directory)

Catalonia and Madrid (both regions experienced a decrease of more than 1%). Since cultural sectors are the core of creative activities and have historically been heavily subsidised by the government, the tremendous cutbacks in public expenditure mean that they have suffered the most; particularly where they used to represent a very significant percentage of the overall contribution of creative sectors.

Urban agglomerations concentrate creative industries in Spain. Creative professionals and companies appreciate both hard and soft factors that the urban arena provides. However, there is a significant difference with regard to their location in the territories: while urban cores accommodate a higher percentage of creative occupations in Madrid and Valencia, Barcelona and Bilbao offer a polycentric approach within their urban regions (Sánchez-Moral et al. 2014; Pareja-Eastaway et al. 2007).

7.4 Promoting the Creative and Knowledge Industries in Spain

7.4.1 The Role of Pathways and Actors

The centrality of creative and knowledge industries in the economic development strategies of Spanish cities after the crisis depends on the uneven pathways of development that these cities have experienced over time. In fact, Spain's uneven and weak process of modernisation has brought about a national urban system with two large metropolitan regions that concentrate an important part of economic activity, and a system of medium-sized cities (more than one million inhabitants) that are regional capitals and have a secondary role in terms of economic development; for example, Valencia, Seville, or Zaragoza. Bilbao, which in terms of population is in this second group of cities, has considerable economic significance and is part of a wider metropolitan region that includes the whole Basque Country. This means that we can find greater density and diversity of economic activities in Madrid and Barcelona, and to a lesser extent in Bilbao, than in Valencia, which means greater possibilities for innovation and creativity (Scott 2008). Nevertheless, we find different historical pathways among cities that influence these possibilities. A short summary of these pathways is necessary to understand the current economic situation of the four case studies.

Being the capital of the state, Madrid's development path has been clearly influenced by being home to the national government and by the decisions taken at the national level. During the industrialisation of Spain (starting in the second third of the nineteenth century), Madrid tended to concentrate administrative services and press industries, with no heavy industrialisation until the 1920s. Nevertheless, together with certain industrial activities there was a powerful services economy linked to the state and the emergence of newspaper and communication industries. Over time, sectors such as cinema and TV, music, the performing arts etc. would grow on the basis of the leadership of the public sector, especially during the Francoist dictatorship.

Conversely, Barcelona developed a powerful local bourgeoisie, which transformed the city into the largest and most relevant industrial centre of Spain. Cultural industries emerged as strong sectors, but they were heavily dependent on private investments, possibly due to the strength of the industrial bourgeoisie. There was an agglomeration of cultural industries, including publishing, advertising, leisure industries (performing arts, music), and media (radio and cinema), and growing public investments in Catalan culture as a means to strengthen the local elites. In many ways, all of these developments, together with the strong efforts of working-class organisations to promote culture and education among workers, transformed the city into the cultural capital of Spain until the 1930s, when the civil war changed this pattern. The victory of Francoism brought about the strong centralisation of cultural industries in Madrid, which took the lead from then on.

At the same time, Bilbao also developed a local middle class, which fostered industrialisation based on heavy industries. Cultural industries played a minor role. In comparison to Catalonia, the development of cultural industries based on Basque culture was weaker, despite the organisation of the bourgeoisie in a nationalist party and the claims for recovering Basque culture in general. In Valencia, industrial development was linked to the modernisation of traditional agricultural activity, and there was a certain cultural movement together with the development of cultural industries, but with little relevance (Pareja-Eastaway and Pradel i Miquel 2014). The development of a local Catalan culture[4] and the links with Barcelona were strong, but during Franco's dictatorship local elites started to become more dependent on Madrid.

From the 1950s onwards, the dictatorship developed a second wave of industrialisation in Spain through a top-down Fordist state policy privileging certain cities. This approach followed policies developed elsewhere in Europe but in a context of a lack of democracy and the strong role of local oligarchic elites. Apart from the obvious choice of Madrid, Barcelona, Bilbao, and Valencia were also candidates for investment and the government launched policies to foster their industrial growth, creating opportunities for the local bourgeoisie. Despite the strong centralism of the regime, local elites tried to develop local initiatives and autonomous city projects, especially in Barcelona, where the bourgeoisie was traditionally strong.

The crisis of the dictatorship came at the same time as the global crisis of Fordism, and the main cities of Spain experienced severe adversity based on growing unemployment, the lack of urban infrastructures and public facilities, and the demise of industrial activity. During the 1970s and the 1980s, alongside the transition to a democratic system, the economic transformation of the country took place. Autonomy for local and regional development policies returned with democracy at the end of the 1970s, and Spanish cities started to develop local strategies for growth in the context of de-industrialisation. Nevertheless, the new

[4]Valencia was conquered by the Aragonese crown and repopulated by Catalans in the Middle Ages, remaining part of the Aragonese Kingdom until its integration into the Spanish state.

multi-level governance framework tended to concentrate competences and resources in newly created regional governments, forcing cities to search for resources and to collaborate with the private sector. Comparing the four cities, we find two different strategies for growth. On the one hand, Barcelona, and to a lesser extent Bilbao, developed strategies based on long-term planning and by creating consensus around large projects for city development. This approach involved integrating private actors and civil society in decision-making mechanisms that guided strategies for growth. On the other hand, since the early 1990s, Madrid and Valencia developed strategies based on a liberal perspective, giving centrality to private actors in economic growth and allowing certain sectors to expand.

During the 1980s, Spanish local governments mainly served to develop infrastructures and facilities for social services, especially in working-class neighbourhoods that did not receive public investments during the dictatorship. Additionally, Barcelona and Madrid started to develop policies to diversify their economy. Madrid, which was already a city with a significant services economy, reinforced this trend with the strong development of cultural institutions and the growth of the financial sector, which was increasingly oriented towards Latin American markets. Barcelona, for its part, started a policy of urban and economic transformation boosted by the opportunity to celebrate the Olympic Games in 1992. The preparation and celebration of the Games led to a strong visitors' industry[5] in the city and allowed for the launching of long-term plans to create the basis for the knowledge economy. This resulted in the consolidation of the role of new economic actors in the city's governance, who tended to collaborate with the City Council in public-private partnerships.

Mirroring the policies developed in Barcelona during the 1990s, Bilbao started a process of urban renewal aimed at diversifying the economy and attracting visitors to the city through the establishment of the Guggenheim Museum and the development of the industrial part of the city to create the conditions for the knowledge economy. In contrast to Barcelona, rather than attracting new activities, Bilbao's efforts were focused on connecting already existing industrial sectors with knowledge-intensive activities. Lastly, since 1991, Valencia has developed a liberal strategy based on the promotion of the real estate sector and the deregulation of land use, bringing about a sharp increase in real estate and speculative actors over traditional agricultural and industrial elites, who had to adapt to the new model.

In 1994, Spain experienced an economic recession resulting from the lack of modernisation of the productive system: after the fall of the iron curtain, Spanish industries became less competitive in European markets in terms of labour costs, largely due to low knowledge intensity. In order to promote economic growth, in 1996, the newly elected conservative government extended the Valencian model to the whole country. The liberalisation of land use led to a speculative bubble in the

[5]We refer here to the tourism industry and the celebration of events and meetings that bring business visitors to the city.

construction sector that generated a decade of economic growth (1997–2007); especially affecting the cities around the Mediterranean coast (Méndez et al. 2012). Despite the uneven effect of the economic model in the different cities, all of them experienced the emergence of construction and real estate sectors as main motors of growth, influencing previous strategies for economic development. To a certain extent, urban renewal in Barcelona and Bilbao, conceived as a means to foster knowledge sectors, became an end in itself, promoting gentrification in certain neighbourhoods without the clear attraction of knowledge-intensive activities. In order to attract investments for the construction sector, Valencia and Madrid used cultural infrastructures and sports events. In Madrid, the financial and banking sector developed private cultural institutions, whereas Valencia created large cultural facilities like the City of Arts and Sciences and venues for the celebration of conferences. These investments not only provided new facilities for Valencia but also contributed to boosting land prices in the whole city and its region, thus attracting new investments.

In short, we find differences in the strategies for economic development in the four cities analysed. Madrid relied largely on being the capital of the country and consolidating the role of the financial sector and the services economy, with the continuous demise of industrial sectors since the 1990s. In this context, the cultural industries grew because of the strong role of public and private actors' investments in them. With the speculative real estate bubble, Madrid's authorities deregulated urban planning and promoted urban expansion and suburbanisation. Valencia followed a similar approach, promoting the deregulation of land, strengthening real estate sectors, and trying to attract visitors and investments to the city in order to foster investments in real estate. Conversely, Bilbao and Barcelona, which had stronger industrial traditions, developed long-term planning strategies to transform their industrial bases into the knowledge-intensive economy. Nevertheless, the development of creative industries was linked largely to regional cultural institutions and local policies to use culture as a factor for economic growth (Rodriguez Morató 2008). Moreover, in Barcelona, the organisation of large events has been increasingly linked to knowledge-intensive sectors. The celebration of events such as the Mobile World Congress has had an impact on the emergence of new businesses and the settlement of IT companies in the city and other companies from creative sectors linked to IT, like video games.

The inclusion of a diverse range of actors and sensibilities (i.e. trade unions, universities, company representatives, municipalities, and so on) in strategic planning in Barcelona since 1992 has resulted in a governance scenario where long-term projects have been possible thanks to an agreed assessment of the priorities and tools needed to address the city's future. On the opposite end of the spectrum, Madrid represents the changing supremacy of actors' interests revealed in consecutive changes in the political colour of the local government, concluding in market forces as the most significant driver, and resulting in difficulties in implementing long-term developments.

The economic crisis in 2008 severely damaged the logics of economic transformation based on a powerful real estate sector, giving way to the emergence of new actors and new forms of organisation. In the economic field, some sectors collapsed whereas new economic activities have gained relevance. The construction sector has declined steadily since 2007, creating huge unemployment and indebtedness in those cities that were more reliant on this economic activity. As a result, both Valencia and Madrid became strongly indebted and unemployment reached its highest peak in history. Barcelona and Bilbao experienced a more limited impact from the crisis in economic terms given the greater diversification of their economies. Both cities continued to generate economic growth during the crisis, mainly through strengthening the tourism sector, but also thanks to the long-term investments in diversification and the existence of knowledge-intensive sectors. In spite of that, both cities, but especially Barcelona, saw a rise in unemployment and social inequalities due to the consolidation of service activities with low added value, like tourism, as main providers of employment (instead of construction and industry which in comparison provided high salaries). The new context for economic development and the thriving relevance of the creative and knowledge sectors question the traditional approach of industrial policies geared towards stimulating economic growth. In this sense, the emergence of new forms of interaction, location, and production among creative and knowledge workers provides an alternative model based on collaboration and cooperation. Cities differ in their provision of adequate contexts for this to happen: for instance, in those cities where these new relationships between companies are gaining weight, facilities for formal and informal networking and knowledge exchange on a transversal basis are growing. As we will see later on, this is more generalised in Madrid, Barcelona, and Valencia.

In the political field, the end of the growth model and the growing social unrest meant that in the municipal elections in May 2015, three of the four cities analysed in this chapter elected municipal governments formed by new political platforms promoting anti-austerity policies, transparency, and citizen participation in policy-making. The electorate has punished local elites that supported the development of growth models based on deregulation and liberalisation, electing more socially-centred governments. Despite the fact that it is still quite early for an assessment, this context seems to present a new opportunity for non-hegemonic economic sectors to develop, as these new governments try to hinder the power of already well-established sectors in the city, mainly the tourism sector.

7.4.2 Policy Approaches: Top-Down, Bottom-Up, and Bottom-Linked

The pathways analysed in the previous sections have created different institutional contexts for creative and knowledge sectors and explain different policy approaches to the promotion of these sectors during the crisis. The four city regions implicitly or explicitly targeted cultural, creative, and knowledge sectors as an alternative to the traditional productive model in Spain but with uneven efforts. As we will see, it is hardly possible to define a clear line between the strategies used by each city as a mixture of approaches can be identified in every case, with differing importance over time. Top-down approaches explicitly support knowledge or cultural sectors in Barcelona or Bilbao; bottom-up initiatives characterise the cultural scene in Valencia and Madrid; and bottom-linked actions are found in the 'Art Factories' programmes in both Madrid and Barcelona.

Thus, we find different conceptions of the role that creative and knowledge industries should play in the overall local economy. In this regard, we again find differences between Madrid and Valencia on the one hand, and Barcelona and Bilbao on the other. As mentioned earlier, since the 1990s, Barcelona and Bilbao have developed projects for urban renewal geared towards creating a physical environment for the knowledge economy. For instance, the 22@Barcelona project, an innovation district initiated in 1998 in a traditional industrial site in Barcelona (El Poblenou), or the *Plan de Renovación de Bilbao y la ría del Nervión* (Renewal of Bilbao and the Nervión estuary). These projects, based on refurbishing the old industrial areas of both cities and populating them with new companies, hoped to create a new source of economic growth after the demise of industry, positioning the cities in regional networks of innovation. These developments were accompanied by the culture-led regeneration of the city centres and the promotion of culture as a source of economic growth. This is the case of the well-known Guggenheim Museum in Bilbao and the large number of initiatives such as the National Theatre, the Auditorium, or the Design Museum in Barcelona, all linked to the promotion of culture. In both cases, the creation of cultural facilities and the revalorisation of existing ones brought increased tourism and fame to the cities. These long-term strategies provided possibilities for growth in certain sectors, especially in knowledge-intensive activities, but at the same time hindered opportunities for growth in other sectors. For instance, the strong emphasis in Barcelona's strategy on knowledge-intensive sectors harmed the emergence of new cultural and creative activities at the beginning of the twenty-first century (Martí-Costa and Pradel i Miquel 2012).

In Valencia and Madrid, short-term, less planned strategies led to the predominance of some sectors. Since the second half of the 1990s, the construction sector and financial activities have played a predominant role in local agendas for growth, and have dominated strategies and economic growth patterns. Despite these sectors being dominant in the national economy and relevant in Barcelona and Bilbao, they were not at the centre of local strategies in these cities, which focused on plans for creative and knowledge sectors as we mentioned above. As the country's capital, Madrid also developed cultural-led regeneration policies, including the fostering of an 'Art Mile' (*Milla del Arte*) in the city centre, taking advantage of the national museums on the *Paseo del Prado* and the creation of cultural centres and facilities. This cultural policy includes the promotion of creative sectors, understanding cultural facilities as a motor for creative activities. At the same time, the city launched bids for the celebration of the Olympic Games in 2012, 2016, and 2020, without any success. Valencia, for its part, also developed large cultural facilities in order to compete and to attract investments, combined with the strong promotion of tourism and the organisation of large events in the city (for instance, participation in the F1 World Championship with the Valencia Street Circuit).

In short, despite similar actions for economic development consisting of urban renewal, the creation of cultural facilities, and the promotion of large events, the main difference between the four cities is the existence or not of long-term strategies for economic growth. The selection of certain sectors linked to the creative and knowledge economy as new cornerstones of economic development was central to these strategies. The most representative case has been Barcelona, which initially identified four clusters associated with the knowledge economy and later added a fifth one based on the creative industries (design). These clusters would be the basis for the growth of economic activity in the 22@Barcelona district. However, this model is not in force anymore and Barcelona will focus from 2015 onwards on a sustainable model of production rather than labelling creative and knowledge industries as a target.

A second consequence of the different institutional contexts in the four cities is the approach to fostering creative and knowledge sectors and the role that these sectors take on in the framework of the economic crisis. Madrid and Barcelona, being the largest cities and with the highest population and economic density, have had a historical role in concentrating cultural industries. Culture-led urban regeneration taking place in the last 20 years has reinforced the role of such industries, creating a market dependent on public demand. The celebration of festivals and large events, together with the creation of networks of public theatres, museums, television channels, and other cultural facilities has allowed the growth of small- and medium-sized companies in creative sectors, especially in Barcelona and Madrid. Nevertheless, the strong impact of the crisis on public administrations has brought with it the demise of creative sectors, strongly dependent on public demand for services. As we have seen, the bursting of the speculative bubble and the arrival of the crisis has meant the demise of creative sectors in these cities, and the emergence of new patterns of organisation. In Barcelona, the gaming industries and app development seem to have an opportunity thanks to the annual celebration of the Mobile World Congress (MWC). Barcelona's bid to celebrate this event was part of its long-term strategy for economic growth based on knowledge, and it is a source of economic activity in the creative sectors. Furthermore, the city has developed the World Mobile Capital programme in collaboration with the MWC organisation to strengthen these sectors. The Mobile World Capital programme fosters the internationalisation of companies and develops collaborations between the health and education sectors and companies working on mobile technologies.

In the context of the crisis, the emergence of new spaces of production, including fab labs, co-working spaces, and the organisation of the makers' movement, can be understood as a response to the new precarious situation. As an example of this trend, co-working spaces in Spain have been growing steadily since the start of the crisis. In 2013, Spain was the third country in terms of the number of co-working spaces after the USA and Germany, according to data from the Global Co-working Census, with 199 spaces (Deskwanted 2013). One hundred and seventy of them are located in Madrid and Barcelona. Despite the fact that private initiative is mainly responsible for the establishment of these spaces, local authorities have also promoted these new forms of work organisation through new policy programmes supporting incubators or start-ups and locating them in local government premises; for instance, the Incubator Programme in Barcelona or Start-up Alcobendas in Madrid. In Barcelona, the impact of the crisis was combined with a change in local government in 2011, bringing a conservative coalition into power. The conservative government (2011–2015) pushed for a clearly liberal approach to economic growth, softening long-term strategies based on knowledge and fostering already growing economic sectors in the city, mainly tourism. In order to promote employment, this government enforced policies for the self-organisation of civil society and the creation of new spaces for innovation, such as fab labs and co-working spaces. In Madrid, this approach was already in motion since the 1990s, with the existence of a network of company incubators, some of them geared towards creative and knowledge sectors. Bilbao, on the other hand, invests in the generation of new initiatives linked to innovation through public and civil society collaboration. Valencia, by contrast, has had a tradition of weak public intervention in economic promotion, giving importance to private actors strongly linked to the construction and tourism sectors. This has given a secondary role to innovation and creativity, but allowed for the emergence of a local creative scene strongly rooted in the territory.

Thus, in the four cities studied, we witness the emergence of bottom-up initiatives to create new forms of production linked to the creative and knowledge sectors. These initiatives are partially the result of increasing precariousness and insecurity in employment, which has brought about new forms of collective organisation and the creation of new institutions not only to share costs but also to exchange knowledge, contacts, and opportunities. This has meant the transformation of small- and medium-sized companies into networks of self-employed people collaborating and, thus, the emergence of spaces to make this collaboration possible. Private actors are responsible for creating these spaces through a purely bottom-up approach, while public authorities are pushing for the creation and institutionalisation of new forms of collaboration.

What we have noted throughout our research is a certain lack of consistency among policies addressing the creative and knowledge sectors and their needs. Thus, flexibility is needed to adapt policies and actions to the new environments and productive formats that characterise these sectors.

7.5 Conclusions

In this chapter, we have analysed the creative and knowledge industries in four large metropolitan cities in Spain through the identification of three dimensions: pathways, actors, and policies. The recognition of particular idiosyncrasies in the creative and knowledge sectors and their own way of confronting the economic crisis enables them to represent a clear alternative for developing new patterns of economic growth. Thus, to understand and enhance their relative importance, this chapter confirms the relevance of the three selected dimensions both at the territorial and the sectoral level. Policies have been strongly attached to the local environment; therefore, the inclusion of territorial characteristics such as pathways and the existing constellation of actors becomes essential for understanding the evolution of these sectors.

Since 2008, the global economic crisis has negatively affected Spain, which has experienced poor rates of economic growth and huge increases in unemployment. The public authorities have seen a powerful cutback in their activities, given their need to fulfil debt and deficit commitments with the rest of Europe. The need to transform the Spanish productive model after the expansion period (2000–2007) requires a reorientation of productive specialisation towards sectors with high added value. Nevertheless, the national government has prioritised measures for immediate economic growth combined with austerity approaches, tending to foster already existing economic sectors. The most important metropolitan agglomerations in the country, i.e. Barcelona, Bilbao, Madrid, and Valencia, have had to confront the challenge of developing new patterns for economic growth in a context of political change. As mentioned, three of the four cities analysed have faced a change in government with new city councils trying to foster new models for production, but they depart from previous economic development models and pathways. Besides, in these new proposals creative and knowledge sectors are not necessarily central elements of the strategy.

The embeddedness of policies and actions in local contexts is a condition for their effectiveness. Our hypothesis in this chapter was that certain significant dimensions contribute to the transformation of the pattern of economic growth in each city: the role of pathways and economic traditions, the complex interaction of existing actors, and the approaches developed to overcome the crisis. We have explored how the territories in question have developed different strategies taking into account these dimensions. The short summary of different development paths in the four cities shows that top-down policies for economic growth developed since the 1990s help to explain differences in the current situation of the cities and their opportunities for developing new economic sectors. Long-term plans for economic transformation towards the creative and knowledge-intensive economy have had an impact in Barcelona and Bilbao, despite their being influenced or biased by the enormous profits gained in the booming real estate sector. More free-market oriented perspectives, such as those in Valencia and to a lesser extent in Madrid, have reinforced the role of hegemonic actors mainly in the construction sector, leading to stronger dependency on them and, with the crisis, the lack of an economic growth model. This context has provided different opportunities for actors in the creative and knowledge industries to become relevant economic actors in their cities.

The cultural sector, core of the creative industries, encounters different channels to flourish and a varied range of policy mechanisms to support it in the four cities. While in some cases a formal perspective is adopted, indeed a particular way of understanding culture, resulting in facilities such as the 'Art Mile' in Madrid or the Culture and Art Palace in Valencia, other forms of cultural activities emerge that are strongly attached to the characteristics of the creative sectors in general, particularly in Barcelona. For different reasons, bottom-up or bottom-linked approaches have emerged in the four cities since the beginning of the crisis, with differing importance. In some cases, they are merely alternative initiatives to the standard labour market to cope with unemployment and precarity. In others, they involve public policy initiatives that discern and gather the needs of the creative and knowledge sectors and respond to the new challenges by providing or facilitating access to new spaces for interaction and creativity.

References

Boix R, Lazzaretti L, Capone F, de Propis L, Sánchez D (2010) The geography of creative industries in Europe: Comparing France, Great Britain, Italy and Spain. Paper presented at Sustainable Regional Growth and Development in the Creative Knowledge Economy – ERSA Congress, 19–23 August 2010, Jonkoping, Sweden

Bontje M, Musterd S, Pelzer P (2011) Inventive city-regions: path dependence and creative knowledge strategies, Urban and regional planning and development series. Ashgate, Aldershot

Camagni R (2005) Economía Urbana. Antoni Bosch, Barcelona

DCMS (1998) Creative industries mapping document. Department for Culture, Media and Sport, London

Deskwanted (2013) Global co-working census. Available at: http://www.zonacoworking.es/wp-content/uploads/2013/02/Global-Coworking-Census-2013.pdf. Accessed 15 Feb 2016

d'Ovidio M, Pradel i Miquel M (2012) Social innovation and institutionalisation in the cognitive-cultural economy: two contrasting experiences from Southern Europe. Cities 33:69–76. doi: http://doi.org/10.1016/j.cities.2012.07.002

Flew T (2010) Toward a cultural economic geography of creative industries and urban development: introduction to the special issue on creative industries and urban development. Inf Soc 26:1–7. doi:10.1080/01972240903562704

Florida R (2002) The rise of the creative class: and how it's transforming work, leisure, community and everyday life. Basic Books, New York (Chapters 1 and 14: pp. 1–17 and pp. 249–266)

García-Montalvo J, Peiró JM (2008) *Análisis de la sobrecualificación y la flexibilidad laboral*. Observatorio de inserción laboral de los jóvenes, IVIE. Available at: http://www.fundacionbancaja.es/archivos/publicaciones/OBSERVATORIOINSERCIONLABORAL.pdf. Accessed 13 Jan 2016

Lange B, Pradel i Miquel M, Garzinov V (2010) New governance, new geographical scales, new institutional settings. In: Musterd S, Murie A (eds) Making competitive cities. Wiley-Blackwell, Oxford, pp 306–326

Lazzaretti L, Boix R, Capone F (2008) Do creative industries cluster? Mapping creative local production systems in Italy and Spain. Ind Innov 15(5):549–567. doi:10.1080/13662710802374161

Le Galès P (2002) European cities: social conflicts and governance. Oxford University Press, New York/Oxford

Martí-Costa M, Pradel i Miquel M (2012) The Knowledge City Against Urban Creativity? Artists' Workshops and Urban Regeneration in Barcelona. Eur Urban Reg Stud 19(1):92–108

Méndez R, Sánchez-Moral S (2011) Spanish cities in the knowledge economy: theoretical debates and empirical evidences. Eur Urban Reg Stud 18(2):136–155

Mendez R, Michelini JJ, Prada J, Tebar J (2012) Economía creativa y desarrollo urbano en España: una aproximación a sus lógicas espaciales. EURE (Santiago) [online] ISSN:0250-7161, 38 (113):5–32. Available at: http://dx.doi.org/10.4067/S0250-71612012000100001. Accessed 4 Nov 2015

Méndez R, Sánchez-Moral S, Malfeito-Gaviro J (2015) Employment changes in knowledge-based industries in large urban areas of Spain: impact of the economic crisis and austerity policies. Environ Plann C Gov Policy 0263774X15614698, first published on November 24, 2015

Musterd S, Kóvacs Z (eds) (2013) Place-making and policies for competitive cities. Wiley-Blackwell, London

Musterd S, Bontje M, Chapain C, Kovacs Z, Murie A (2007) Accommodating creative knowledge. A literature review from a European perspective. ACRE Report 1. Amsterdam: AMIDSt. ISBN:978-90-78862-01-7

Novy J, Colomb C (2012) Struggling for the right to the (creative) City in Berlin and Hamburg: new urban social movements, new "spaces of hope"? Int J Urban Reg Res 37(5):1816–1838

Pareja-Eastaway M, Turmo J, Pradel i Miquel M, Garcia Ferrando L, Simó Solsona M (2007) The city of marvels? Multiple endeavours towards competitiveness in Barcelona. Pathways to creative and knowledge-based regions. ACRE Report WP2.2. Amsterdam: AMIDSt. ISBN:978-90-75246-56-8

Pareja-Eastaway M, Turmo J (2013) La necesaria transformación del modelo productivo en España: el papel del territorio. Documents d'Anàlisi Geogràfica 59(3):455–479

Pareja-Eastaway M, Pradel i Miquel M (2014) Towards the creative and knowledge economies: analysing diverse pathways in Spanish cities. Eur Plan Stud 23(12):1–19. doi:http://doi.org/10.1080/09654313.2014.988018

Pradel i Miquel M (2016) Kiezkulturnetz vs. Kreativequarier: social innovation and economic development in two neighbourhoods of Berlin. City Cult Soc (Forthcoming)

Pratt A (2012) A world turned upside down: the cultural economy, cities and the new austerity. Smart, creative, sustainable, inclusive: territorial development strategies in the age of austerity. Regional studies association winter conference, London

Pratt A, Hutton T (2013) Reconceptualising the relationship between the creative economy and the city: learning from the financial crisis. Cities 33:86–95

Recio A (2010) Capitalismo español: La inevitable crisis de un modelo insostenible. *Revista de Economía Crítica* 9 (first semester), pp 198–222

Rodriguez Morató A (2008) La emergencia de una capital cultural europea. In: Degen M, García M (eds) La metaciudad: Barcelona. Transformación de una metrópolis, Anthropos, Barcelona, pp 45–64

Sánchez-Moral S, Méndez R, Arellano A (2014) Creative economy and employment quality in large urban areas in Spain. Urban Geogr 35(2):264–289

Scott AJ (1997) The cultural economy of cities. Int J Urban Reg Res 21(2):323–339. doi:http://doi.org/10.1111/1468-2427.00075

Scott AJ (2008) Social economy of the metropolis: cognitive-cultural capitalism and the global resurgence of cities. Oxford University Press, Oxford

Sennett R (2008) The craftsman. Yale University Press, London. Retrieved from http://www.amazon.co.uk/dp/0141022094

Storper M (2013) Keys to the city: how economics, institutions, social interaction, and politics shape development. Princeton University Press, Princeton. Retrieved from http://press.princeton.edu/titles/10022.html

Chapter 8
Creative Industries in the Polish Economy: Growth and Operating Conditions

Rafał Kasprzak

8.1 The Creative Sector in Poland

8.1.1 Classification of the Sector in the Polish Context

The boundaries of the creative sector as an economic sphere are difficult to identify. To set the stage for our further analyses, we must at least briefly discuss the meaning and designation of the term, which will serve as a foundation for the analyses of geographical differences and as a baseline for specifying the key determinants of the creative sector's development in the Polish economy during the years of the crisis and thereafter. The scientific literature devoted to creative industries proposes two principal methods to identify those organisations that belong to the creative sector.

In the first method, the principal criterion focuses on the sector's products, which should be protected under copyright law as their key feature is that they constitute intellectual property. Despite the fact that the 'product' of the creative sector has a few other important features, such as cultural value, symbolism, communication capacity, and external effects, they are assumed by definition to yield intellectual property, which can be secured and licensed. Such a perspective is analysed in depth in reports published by *The International Intellectual Property Alliance* (see their report 'Copyright Industries in the U.S. Economy'). Features of creative sector products are also elaborated on by, *inter alia*, Throsby (2010), Samuelson (1954), Musgrave (1959), Marshall (1920), and Pigou (1920).

The second method of identifying organisations within the creative sector examines processes within an organisation and determines to what extent they are creative or consist in delivering certain experiences or creating a specific element of

---

R. Kasprzak (✉)
Warsaw School of Economics, Institute of Management, Warsaw, Poland
e-mail: rafal.kasprzak@sgh.waw.pl

© Springer International Publishing AG 2017

C. Chapain, T. Stryjakiewicz (eds.), *Creative Industries in Europe*,
DOI 10.1007/978-3-319-56497-5_8

symbolic culture. In this approach, a given group of economic activities is usually identified by assuming, sometimes arbitrarily, that they are creative by nature. This concept of the creative sector was popularised by Howkins (2001), who proposed a list of 15 industries which can be identified as creative. The *UK Department for Media Culture and Sport* uses a similar, if not identical, approach.

The above methods of classification have led to several definitions to describe the creative sector. In Poland, the most popular definition is the one proposed by the *UK Department for Culture, Media and Sport* (Department for Culture, Media and Sport 2001), which defines creative industries as areas of economic activity which have their origin in individual creativity and whose products involve intellectual property rights. This definition lists individual fields of activity that form parts of these industries, such as advertising and marketing; architecture; crafts; design; film, TV, video, radio, and photography; IT; publishing; museums, galleries, and libraries; and music and performing and visual arts. This classification method allows for a detailed analysis of their national economic impacts as well as comparative studies between regions and states. Using this as a basis, when trying to define the sector according to the realities of the Polish economy, we may identify it as an area of services focused on the improvement of human capital and covering the generation and commercialisation of cultural products, involving different organisational forms of economic activities, in the following sub-sectors:

- Arts and crafts, including visual arts, performing arts, as well as national heritage, libraries, and archives
- Creative production, including computer programming, publishing, film and TV production, as well as radio and music production
- Creative services, including fashion and design, advertising and related activities, architecture and interior design[1]

The creative sector understood in this way can be divided into several areas with individual codes of classification of their economic activities according to NACE (see Appendix). The creative sector combines different fields of activity and various objectives. One of its basic characteristics is an important internal differentiation with respect to organisational forms, goals, and location. When analysing the variety of organisational forms we may observe their different organisational structures, such as enterprises (e.g. sole proprietorships, partnerships, limited liability companies, etc.), public institutions (e.g. administrative units at the central, regional or county levels), and NGOs (e.g. foundations, associations, economic organisations). This highly diversified pattern of organisational forms is both based on and translates directly into a variety of objectives. The creative sector seems to be a unique sector in the contemporary Polish economy, in which profit-

[1]This typology is one of many attempts to scale and analyse the creative sector. Although it is the most popular approach in Poland, there are also other typologies. A complex presentation of typologies which can be used for creative sector analysis is presented in, *inter alia*, KEA European Affairs (2006).

oriented operators coexist and cooperate with entities whose performance is measured by indices other than profit (e.g. ensuring access to culture and national heritage).

The above-presented different aspects, i.e. organisational forms, goals and location, pose key challenges to economic policy planning, which results in a coordination dilemma. Each organisational form requires certain legal solutions for its establishment, management, and monitoring, which requires a lot of cross-sectoral cooperation in assigning and coordinating a strategy for this sector.[2] In the Polish economy the coordination dilemma can be easily observed in financing strategies[3] and in promotional activities, which use the various possibilities offered by the sector.[4]

8.1.2 Spatial Differences in the Sector

The data on the number of the creative sector entities, as contained in the REGON register[5] from the Central Statistical Office of Poland (Table 8.1), allow us to analyse their spatial distribution and development dynamics (growth ratios). However, the REGON database has some significant disadvantages which must be taken into consideration. The first concerns the moment when the data is entered into the register. In most cases, this takes place when enterprises are first registered and, unfortunately, subsequent updating is relatively rare. The second disadvantage lies in the fact that an entity may be formally registered, but in fact does not actively pursue any economic operations in its field of activity. However, despite the fact that the list of registered entities by their NACE codes sometimes does not reflect their actual extent of business activity, REGON is still the only database in Poland

[2]In the Republic of Poland, the Ministry of Culture and National Heritage and the Ministry of Economy are assigned responsibility for the coordination of economic policy towards the creative sector. The Ministry of Culture and National Heritage focuses on the coordination of the governmental part of the sector, as it provides funding for cultural organisations and creates the legal environment for most of them. The Ministry of Economy develops legal regulations and supporting instruments addressed to private enterprises in the sector.

[3]Key sources are public funds distributed by governmental organisations (which, however, are managed by governmental bodies and local governments), private funds distributed by NGOs (many organisations at both the national and local levels), and other private funds (very rare in the Polish economy). The majority of funding is distributed via public organisations, which requires a lot of coordination effort and is difficult to implement.

[4]Many counties are strengthening their promotional efforts, using the possibilities offered by the creative sector (i.e. cultural events, great artists, creative districts in cities, amusement parks, etc.). However, in Poland there are many institutions responsible for international promotion, which makes the task of coordination a real challenge.

[5]The REGON register (National Official Business Register) is a continuously actualised set of information on entities of the national economy. It was established in 1995 and is managed by the Central Statistical Office of Poland.

Table 8.1 Number of creative sector (CS) entities in the REGON database from 2009 to 2013 by region (voivodeship)

Area	2009	2010	2011	2012	2013	Growth ratio in the number of CS entities (2013/2009)	Growth ratio in the number of all entities in the national economy (2013/2009)
Poland	135,339	144,042	141,797	141,875	144,236	106.57	108.75
Voivodeship (NUTS 2 Region):							
Dolnoslaskie	10,000	10,776	10,866	10,983	11,054	110.54	109.71
Kujawsko-Pomorskie	5663	5851	5747	5691	5807	102.54	105.07
Lubelskie	5252	5483	5216	5139	5283	100.59	108.70
Lubuskie	2930	3086	2956	2885	2852	97.34	108.32
Lodzkie	7673	8307	8114	8171	8216	107.08	107.80
Malopolskie	12,015	13,002	13,029	13,116	132,19	110.02	111.80
Mazowieckie	34,481	36,289	36,015	36,384	37,456	108.63	112.11
Opolskie	3259	3510	3321	3182	3214	98.62	104.77
Podkarpackie	4096	4523	4481	4458	4580	111.82	110.09
Podlaskie	2827	2976	2995	2987	2981	105.45	107.75
Pomorskie	7902	8416	8266	8249	8422	106.58	109.04
Slaskie	14,122	15,331	14,891	14,832	14,977	106.05	106.91
Swietokrzyskie	3054	3199	3095	3082	3036	99.41	103.98
Warminsko-Mazurskie	3226	3449	3302	3245	3299	102.26	106.45
Wielkopolskie	12,627	13,316	13,229	13,409	13,688	108.40	110.72
Zachodniopomorskie	6212	6528	6274	6062	6152	99.03	102.09

Source: REGON register, Central Statistical Office of Poland

providing information about the number of organisations in the creative sector which are located in the sixteen Polish administrative regions.[6]

As can be seen in Table 8.1, from 2009 to 2013, the growth ratio of the creative sector entities in Poland was slightly smaller than the growth ratio of all entities in the national economy, as registered in the REGON database. The regions with the strongest growth over the analysed period (together with the major city(ies) in each region) were Malopolskie (Krakow), Mazowieckie (Warsaw), and then Podkarpackie (Rzeszow) and Dolnoslaskie (Wroclaw). Decreasing ratios were observed in the Lubuskie and Opolskie regions, which face significant structural problems with their economies. Lubuskie (Gorzow Wielkopolski, Zielona Góra) is located in the western part of Poland and faces the problem of a shrinking economy and migration of its citizens. Opolskie (Opole) is located between two economically strong Polish regions – Slaskie and Dolnoslaskie – which impacts negatively on the migration of its citizens to other Polish regions and also to Germany, as a significant part of its population has dual citizenship.

The creative sector's entities are unequally distributed over the regions, as reflected in the statistical indicators presented in Table 8.2.

As far as the number of creative sector entities is concerned, the following regions (voivodeships) have a dominant position, Mazowieckie, Slaskie, Wielkopolskie, Malopolskie, and Dolnoslaskie, which surely is correlated with the size and prospects of their regional economies. This could be explained by the fact that these regions are the sites of Poland's major metropolises, which seem to operate as strong hubs for the location of creative sector entities. The most attractive location is the capital city of Poland – Warsaw (i.e. Mazowieckie region). This region has by far the highest number of creative sector entities located therein in comparison with the other areas. In the city of Warsaw, the numbers of registered creative sector entities were in 2009, 23,036, and in 2013, 25,285. In the year 2013 as many as 17.53% of all creative sector entities in Poland were located in the city of Warsaw. In contrast, the lowest numbers of creative sector entities were observed in cities and regions located in the eastern part of Poland, i.e. the Warminsko-Mazurskie and Podlaskie regions.

Using a comparison of growth ratios, it can be seen that this particular sector of the Polish economy has been growing (in terms of number of registered entities) slower than the growth rate of other entities in the national economy, i.e. 106.57 vs. 108.75 from 2009 to 2013. There are two exceptions to this – the Podkarpackie (Rzeszow) and Dolnoslaskie (Wroclaw) voivodeships.

---

[6]Following changes in the administrative division in Poland launched in 1999, the Republic of Poland consists of 16 regions (called voivodeship). They are listed here together with the main city (ies) falling within their jurisdiction: Dolnoslaskie (Wroclaw); Kujawsko-Pomorskie (Torun and Bydgoszcz); Lubelskie (Lublin); Lubuskie (Gorzow Wielkopolski and Zielona Gora); Lodzkie (Lodz); Malopolskie (Krakow); Mazowieckie (Warsaw); Opolskie (Opole); Podkarpackie (Rzeszow); Podlaskie (Bialystok); Pomorskie (the Tri-City of Gdansk, Gdynia, and Sopot); Slaskie (Katowice); Swietokrzyskie (Kielce); Warminsko-mazurskie (Olsztyn); Wielkopolskie (Poznan); and Zachodniopomorskie (Szczecin).

Table 8.2 Statistical indicators of creative sector entities registered in 2013 by region (voivodeship)

Area	Creative industry entities			
	registered in 2013	per 10,000 citizens	as a % of all entities registered	per 10,000 m^2
Poland	144,236	37.47	3.54	4613
Dolnoslaskie	11,054	37.99	3.18	5542
Kujawsko-Pomorskie	5807	27.75	3.04	3231
Lubelskie	5283	24.50	3.11	2103
Lubuskie	2852	27.92	2.60	2039
Lodzkie	8216	32.69	3.45	4510
Malopolskie	13,219	39.34	3.77	8706
Mazowieckie	37,456	70.45	5.17	10,534
Opolskie	3214	32.00	3.21	3415
Podkarpackie	4580	21.51	2.87	2566
Podlaskie	2981	24.95	3.09	1477
Pomorskie	8422	36.68	3.10	4600
Slaskie	14,977	32.56	3.25	12,144
Swietokrzyskie	3036	23.94	2.76	2592
Warminsko-Mazurskie	3299	22.80	2.70	1365
Wielkopolskie	13,688	39.48	3.44	4589
Zachodniopomorskie	6152	35.79	2.80	2687

Source: REGON register, central statistical office of Poland

Similar economic and spatial differences can be observed in terms of sub-sector analysis. As mentioned earlier, the creative sectors can be divided into three main sub-sectors:

- Arts and crafts
- Creative production
- Creative services

Changes in the number of registered entities for each sub-sector are presented in Table 8.3.

As can be seen, sub-sector growth ratios were strongly diversified over the period 2009 to 2013. First, the sub-sector of arts and crafts displays an overall decline, especially due to the decrease in the number of visual arts entities, not compensated for by the stable evolution in the national heritage, libraries and archives entities (which was due to their state-owned status), or the slight growth in performing arts entities. Spatially, this sub-sector tends to be equally distributed among regions, with slightly higher numbers in metropolises.

In contrast, interesting trends can be observed in the creative production sub-sector. The number of entities in the programming branch, which in the NACE code refers to the gaming industry, more than doubled from 2009 to 2013.

Table 8.3 Number of registered entities in the creative sub-sectors in Poland from 2009 to 2013

Activity/year	2009	2010	2011	2012	2013	Growth ratio of the number of entities in CS sub-sectors (2013/2009)
Sub-sector of arts and crafts	60,654	60,494	56,542	55,192	55,302	91.18
Visual arts	47,768	46,959	43,322	42,069	41,937	87.79
Performing arts	9824	10,475	10,199	10,075	10,292	104.76
National heritage, libraries and archives	3062	3060	3021	3048	3073	100.36
Sub-sector of creative production	20,113	21,596	20,984	20,774	20,809	103.46
Programming	165	282	329	374	418	253.33
Publishing	10,486	11,190	10,868	10,765	10,605	101.13
Film and TV production	7462	7842	7512	7391	7526	100.86
Radio and music production	2000	2282	2275	2244	2260	113.00
Sub-sector of creative services	54,572	61,952	64,271	65,909	68,125	124.84
Fashion and design	5359	7006	7639	8249	9322	173.95
Advertising and related activities	31,232	33,648	34,443	35,244	36,145	115.73
Architecture and interior design	17,981	21,298	22,189	22,416	22,658	126.01

Source: REGON register, central statistical office of Poland

This rapid growth has already been noticed by policy-makers in Poland, and the gaming industry has been defined as a branch of the Polish economy with one of the best prospects.[7] The relative stable numbers in the film and TV production and publishing branches mean that these branches have probably reached their peak and some mergers can be expected.

The strongest increase in the number of entities was observed in the creative services sub-sectors – up almost 25% from 2009 to 2013. The fashion and design industry has registered particularly strong growth, and almost all metropolises in Poland identified this branch as a prospective opportunity for their local economy.

[7]The gaming industry was enrolled as a national scientific specialisation, which means that this group of activities has a greater possibility to raise public funds for research and development. Moreover, the Ministry of the Economy and some regional authorities introduced tailored grant schemes to foster their development. The list of National Scientific Specializations was accepted by the Polish government on 8 April of 2004; see Ministry of the Economy (2014).

Despite the financial crisis, the advertising and related activities branch continued to grow.

While creative sector entities are an important part of the economy, nonetheless their average share in the number of organisations in the national economy was relatively small (3.54% in 2013). This indicator is significantly differentiated from a regional point of view. The city of Warsaw has the highest share of creative sector entities in its regional economy (6.96%), while the other metropolises with a significant share of creative sector entities include Krakow (5.56%), Poznan (5.10%), Wroclaw (4.80%), the Tri-City of Gdansk-Sopot-Gdynia (4.41%), and Lodz (4.11%).

The analysis of regional differences based on the data from the REGON register allows us to draw some interesting conclusions. The creative sector in Poland is strongly diversified territorially. Its operators are concentrated mostly in big cities in those voivodeships (regions) where the economic situation is favourable. The city of Warsaw deserves special attention with its very high concentration of creative sector operators. There are many reasons why Warsaw is more attractive for the sector than other cities, including purchasing power of households; the 'big city effect', which refers to the high level of demand for creative sector's products; and the 'statistical effect' of headquarters' location, which means that despite the area where their activities are carried out, companies have a preference for locating their headquarters in Warsaw; thus all their statistical reporting refers to this region.

Altogether, the above-listed metropolitan areas accounted (statistically) for almost 35% of creative sector entities in Poland over the period 2009–2013.

8.2 Changes in the Creative Sector Environment in Poland: An Analysis Based on the IDEA Model

This section proposes an analysis of the various dimensions which affect the creative sector in Poland. It uses a model developed by the present author (Kasprzak 2013a), based on a research into how national and local authorities are developing and introducing economic policies towards the creative sector in regions of Poland and in other European countries. The IDEA model identifies four key directions for economic policy towards the creative sector: internationalisation, demand, entrepreneurship, and availability. Following a brief presentation of the model (Sect. 2.1), the following Sects. (2.2, 2.3, 2.4, and 2.5) show how these dimensions are being shaped in Poland.

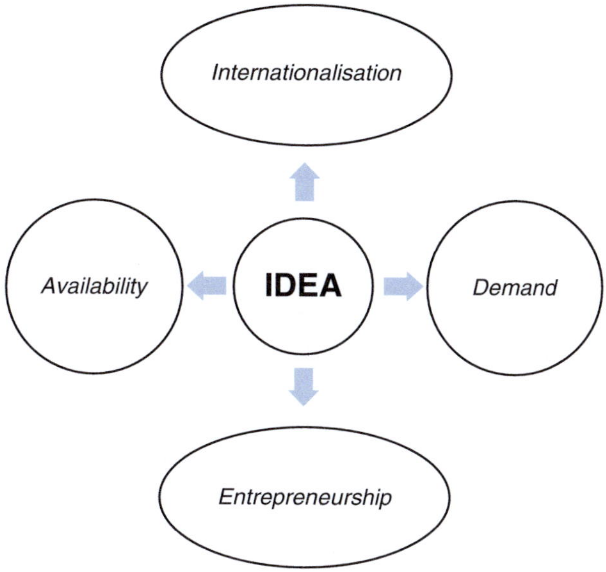

Fig. 8.1 IDEA model: determinants of the growth of the creative sector in the national economy (Source: R. Kasprzak, 2013. Przemysly kreatywne w Polsee – uwarunkowania i perspektywy (Creative industries in Poland: determinants and prospects). Kamon Consulting Warsaw, p. 149)

8.2.1 Introduction to the IDEA Model

Various academic research suggests that the growth of the creative sector requires a series of coordinated active policies and initiatives, with a view towards supporting its growth (see Adamiak et al. 2013; Demel et al. 2004; Fesel and Sondermann 2007; Gibbon 2011; Grochowski 2010; Grochowski et al. 2009, 2012; Reis 2008; Senate Department for Urban Development 2008). To understand the requirements of creative sector entities from the economic policy point of view, it should be mentioned that the presented analysis shows the creative sector as a whole to have similar expectations – see Fig. 8.1. Each of these areas should be translated into sectoral activities on the part of both central and local government administrations. These four fundamental growth determinants are an integral part of the IDEA model, which identifies the key directions for public policy to support the growth of the creative sector.

Internationalisation is an important developmental factor in the creative sector as the supply of products often exceeds the absorption capacity of the local (or national) market. Efficient solutions conducive to the internationalisation of production outputs would surely contribute to the growth of the sector. It is important to note that the specificity of the sector consists in numerous international contacts and in ignoring, to the extent possible, administrative borders in collaboration, e.g. when setting up project teams or working on concrete projects. The state of today's technology enables teams to work around the clock on a project and the various and rapidly expanding types of communications intensify this cooperation even further. Clearly this technology makes it possible for operators in this field to act globally.

However, while in the context of establishing collaboration barriers to communication practically do not exist anymore, as communications technology allows persons to work with anyone worldwide, the situation becomes more complicated in the further stages of a project's execution, especially in relation to administrative requirements. The growth of the creative sector calls for skilful support to be offered to its operators in the internationalisation of their products. This support should cover the following areas:

- Intellectual property protection
- Exports of products of the creative sector
- Promotion of products of the creative sector in overseas markets
- Building networks of international cooperation links

The institutions responsible for supporting the sector should not only focus on the development of ways to provide the kind of support listed above, but more importantly demonstrate to the creative sector how its operators can apply for such support and create incentives to use these instruments.

Demand is the second growth factor in the creative sector. A wide range of activities need to be undertaken to increase the demand for the products of creative industries. Demand should be stimulated via short-term actions (e.g. sponsorship schemes which generate demand for creative industries' products) and long-term activities (e.g. educational activities addressed to altering the patterns of household consumption). The important actors in the area of sponsorship in Poland are the Ministry of Culture and National Heritage, together with those national organisations and local authorities which are responsible, at least in part, for various sponsorship schemes. Households are also important components of demand and they should be encouraged to increase the amounts allocated in their budgets for purchase of the products offered by creative industries.

Entrepreneurship is the third element of support actions which should be introduced by policy-makers with respect to creative sector industries. The creative sector, founded on unique and individual effort, requires a uniform operating framework that would favour its growth. The most important issues in this respect include the facilitation of starting up a business, and easing the multitude of taxes and liabilities with respect to social insurance. From the business point of view, primary importance is attached to improved efficiency of the tax system, which could be achieved by streamlining the rules for calculating taxes, especially VAT (the tax on products and services). Other important areas include encouragement of the establishment of new businesses and strengthening the entrepreneurial spirit among those who operate within the sector.

Availability is the fourth key component in designing a coherent economic policy towards the creative sector. In this context it is necessary to provide three fundamental types of resources which are critical for the creative sector to grow, i.e. capital, human resources, and access to economic and subject area knowledge.

The introduction of a coherent economic policy towards the creative sector according to the IDEA model requires intensive cooperation between national, regional and municipal (or metropolitan) authorities, as each of them has different tools and possibilities. National authorities should focus on the coordination of laws and providing support for regional policies, whereas regional and municipal (metropolitan) authorities should efficiently allocate public support to strengthen each dimension of the IDEA policy model.

The IDEA model can be a useful tool not only for designing a coherent economic policy addressed to the creative sector, but can also aid in identification of the key changes in the sector's environment, i.e. in this study in relation to Poland in the post-crisis period. This aspect is explained, with respect to each determinant, in the subsections below.

8.2.2 Key Changes in the Sphere of Internationalisation

During the post-crisis period in the Polish economy several activities designed to support the creative sector in the sphere of *internationalisation* could be observed. They are presented in Fig. 8.2.

The internal differences in Poland's creative sector industries intensify the problems connected with coordination of the activities pursued in order to support their internationalisation. Such support is offered mainly by the following institutions:

- The Ministry of Culture and National Heritage, Department of Foreign Cooperation, which is responsible for funding projects whose main objective is to promote internationally the most valuable phenomena in Polish culture. The Ministry supports important projects designed to present Polish art in a comprehensive and versatile way and promotes the accomplishments of eminent Polish artists during prestigious global cultural events.[8]
- The Adam Mickiewicz Institute, a state-owned art institution entrusted with the task of promoting Polish culture. During the years 2001–2013, the Adam Mickiewicz Institute organised promotional projects in 26 countries including, *inter alia*, the United Kingdom, Russia, Israel, the Benelux countries, Spain, Austria, Sweden, France, Germany, Ukraine, Lithuania, as well as Algeria, Morocco, India, and China. So far, the projects organised and initiated have included over 4000 cultural events presented by the Institute to 40,000,000 viewers in three continents.

The above institutions focus their activities mostly on the so-called high culture. The remaining part of creative sector activities are promoted internationally by other institutions, such as:

[8]Based on the information available on the website of the Ministry of Culture and National Heritage, www.mkidn.gov.pl (22.07.2015).

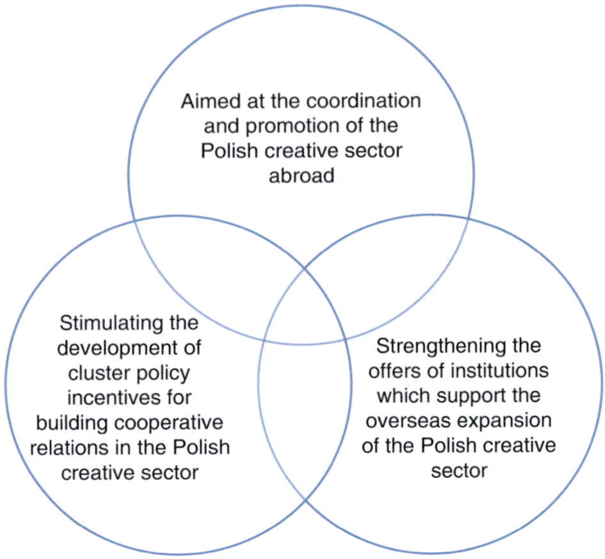

Fig. 8.2 Activities in the sphere of internationalisation (Source: Author's studies)

- The Ministry of Foreign Affairs, which becomes involved in promotional activities if and when authorised to do so by the Prime Minister
- The Polish Tourist Organisation
- The Polish Information and Foreign Investment Agency
- The Ministry of Economy
- Regional Development Agencies
- Chambers of Commerce and associations

We can see clearly that the promotion of the activities of the creative sector is decentralised in Poland, and the coordinating competences are duplicated by a variety of organisations. Undoubtedly this is also an effect of the specificity of the creative sector and the wide spectrum of its operations. The discussion concerning the promotion of products offered by the Polish creative sector increasingly often highlights the need to elaborate mechanisms which would coordinate this decentralised approach. Such efforts are more and more frequently undertaken by local authorities in Poland, especially in metropolises, which are ever more actively striving to develop coherent promotion mechanisms for the sector by using public funds available under the EU Structural Funds.

Since 2004, Poland, as a European Union Member State, has been one of the major beneficiaries of the cohesion policy, principally due to the lagging behind of its economy and various structural problems. Resources from the cohesion policy are used by local authorities and the central government to initiate many programmes aimed at supporting the internationalisation of Polish enterprises and covering expenses connected with: drafting an international expansion strategy; printing foreign promotional and exhibition materials; establishing collaboration with foreign partners; and participation in economic missions and fairs. Enterprises from the creative sector often benefit from such programmes and schemes, as they are able to offer innovative products and their operations easily fit into regional and national innovation or intelligent specialisation strategies. It seems that the support systems developed by the local and central administration authorities can greatly facilitate the establishment of cooperation links between entities in the creative sector.

Increasing and supporting the development of clusters in Poland is another important support element for the internationalisation of the creative sector in the post-financial-crisis period. A classically defined cluster (Porter 2001, p. 246) is a geographical concentration of interlinked firms, specialised suppliers, service providers, and firms from related sectors, together with institutions connected with them (e.g. universities, standardisation bodies, and industrial associations), located in a specific area where they can not only compete with one another but also cooperate.

Thus clusters of creative industries can be perceived as a geographical grouping of enterprises and other institutions from the creative sector. However, due to the importance of creativity in these industries, such clusters also mean the grouping together of talented people, not just institutions (UN-UNCTAD 2010, p. 75). Hence the crucial attributes of a cluster include, on the one hand, the presence of a certain economic potential understood in terms of geographical concentration of operators (enterprises, other organisations, or individuals) from the same or related industry, and on the other hand, the joint developmental efforts of these operators.

The emergence of cluster initiatives in the Polish economy is closely linked to the availability of grants earmarked for building cluster cooperation relationships and for funding cluster operators. According to the PARP Cluster Map,[9] there were 192 clusters in Poland in the year 2015, the majority of them located in the Mazowieckie and Slaskie regions, i.e. mainly in Warsaw and Katowice. In-cluster relations may be vital for enhancing the potential of the creative sector to expand internationally, as the majority of them underline the importance of international expansion for their members. Hence stimulating the development of clusters may significantly contribute to the growth of the creative sector. Cluster policy in Poland is focused on enhancing innovation and improving the competitiveness of the Polish economy by striving to attain the following objectives (Dzierżanowski 2012, p. 23):

- To stimulate internal interactions, knowledge transfers and cooperation, and the development of the necessary human capital
- To increase the external networking of clusters and their operators, especially on the inter-sectoral and international levels
- To strengthen common and integrated strategic planning within clusters

[9]The PARP Cluster Map is a tool, available online, which provides up-to-date basic data about clusters operating in Poland, received directly from their coordinators. Website: http://www.pi.gov.pl/PARP/data/klastry/index_en.html (27.07.2015).

- To increase the number of innovative goods and services offered in the domestic and international markets by enterprises and operators/members of clusters, which translates into, *inter alia*, increased exports
- To encourage private investment in clusters in order to increase the number of start-ups and stimulate the inflow of foreign investment, as well as enhance private outlays on R&D and innovation
- To develop business environment organisations (e.g. educational, scientific, technology parks, technology transfer centres, etc.) and better adjust their offers to the needs of enterprises in the cluster
- To improve the effectiveness of public investments through their concentration and the synergy between policies and support instruments (e.g. in infrastructure development, human resources, R&D, promotion, internationalisation, etc.)

As the pro-cluster policy is still under development, it is very difficult to evaluate its effects on clusters themselves, especially those which group entities from the creative sector.

8.2.3 Key Changes in the Sphere of Demand

The activities supporting the creative sector in the sphere of *demand* are presented in Fig. 8.3.

Stimulating demand for the products of the creative sector is an important part of economic policy, which directly translates into mechanisms which provide funding for activities leading to the creation of cultural products. Funding creative industries from public resources focuses mainly on those institutions of culture responsible for such activities as the creation, promotion/dissemination, and protection of culture. Changes in the funding of creative industries are aimed at (Głowacki et al. 2009):

- Introducing, in particular in the sector of subsidised cultural activities, mechanisms (instruments) that would favour rational, effective, and fair management of public resources (regulations in the area of public procurement and public finance)
- Changing the scope of responsibilities of public administration vis-à-vis the organisation and funding of culture
- Implementing 'new' solutions in the funding, organisation, and management of cultural institutions (decentralisation of management of cultural institutions and giving them more independence, involvement of private entities in performing and delivering tasks connected with arts and culture, developing an appropriate legal framework for public patronage and the sponsoring of culture, etc.)

The differentiated intensity of financial support, together with the possibility of financing cultural institutions at the central level under various schemes run by the

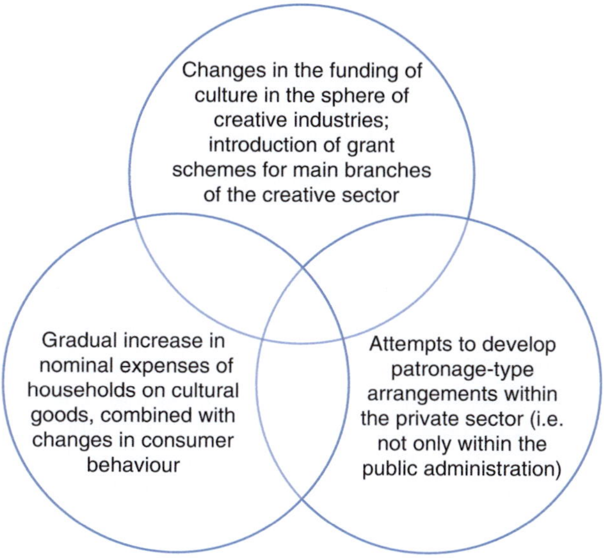

Fig. 8.3 Changes in the sphere of demand (Source: Author's studies)

Minister of Culture and National Heritage[10] translates, in practical terms, into varying degrees of their security and differing amounts being channelled to them depending on who they report to. In order to stabilise their financial standing, cultural institutions often try to ensure for themselves a relatively constant and lasting support from local authorities or from the Ministry. Thus the overlapping competences of the local authorities and the Ministry, largely due to inconsistent legal solutions, are an important issue in the system of financing cultural institutions and currently blur the financing of culture. Changes to the system proposed, *inter alia*, during the Congress of Culture in 2009 have been opposed by certain circles in the sector, which are unwilling to give up ministerial support for their activities.

Clearly the key dilemma connected with the structure of financing the system of culture is the issue of the financial efficiency of culture and the institutions established for its dissemination. The Ministry of Culture and National Heritage, as the constitutional guardian of the principle of the accessibility of culture and the protection of its products (national heritage), must develop financial solutions by which entities in the creative sector may be subsidised under the umbrella of access to culture. Public funding is justified mainly by the generally acknowledged fact that some selected areas of creative industries are unable to self-finance themselves.

[10]Every year the Minister of Culture and National Heritage announces a call for proposals for grant schemes, addressed to various organisations from the creative sector. Grant schemes finance various cultural and creative activities introduced by the sector which are non-profit oriented and have a low possibility of raising private support. Grant schemes are usually branch-oriented (i.e. cultural education, literature, artistic events) and in certain years may be addressed to support a sector celebrating the achievements of key persons from the Polish culture (e.g. the 2010 Chopin anniversary and the 2013 Lutosławski anniversary).

The financial model worked out by the Ministry of Culture and National Heritage assumes that the public sector is the principal, but not the only, patron of culture. While private sponsorship is considered as an additional source, not as an alternative, to the public funding of culture, nonetheless incentives are being introduced in Poland to combine public and private resources. The Minister of Culture was allocated a portion of the central budget for distribution in line with the Budget Law adopted for a particular year and with the Act on Public Finance. Resources earmarked for cultural institutions are also available from local authorities. Culture-related projects may also be financed from international sources, such as the funds of the European Union or the United Nations (World Heritage Fund).

The availability of public resources via several governmental levels (the Ministry, local authorities, EU funds) unfortunately adds to the inconsistency and complexity between what happens at the central (Ministry of Culture and National Heritage) and local levels (units of local authorities responsible for the operation of cultural institutions). In order to resolve the problem, effective coordination mechanisms between the central and regional levels are needed. At present, the Minister has no effective instruments to impact culture in the regions. As a result, the development of a coherent programme of cultural policy constitutes a real challenge. In addition, it must be kept in mind that the system of public funding focuses above all on public cultural institutions, and offers only a small amount of support to institutions from the creative industries which are not owned by the state or local authorities. However, grants are available to finance the activities of the creative sector pursued with a view towards offering products for consumption. They constitute one of the vital elements in building demand for products of creative businesses and encouraging increased household expenditures on the purchase of cultural products and services.

The Polish public statistics regime measures expenditures on the purchase of cultural products and services based mostly on the household budget survey conducted by the Central Statistical Office of Poland.[11] Expenditures on culture and leisure, which include the purchases of products manufactured by creative industries, is an aggregated group of expenditures composed of:

1. Expenditures on audiovisual, photo, and IT equipment
2. Expenditures on other fixed assets connected with culture and leisure
3. Expenditures on other leisure-related equipment, gardening products, and pets
4. Expenditures on culture and leisure services
5. Expenditures on newspapers, magazines, books, stationery, and articles for drawing and painting
6. Expenditures on organised tourism

The analysis of household expenditure, if conducted to estimate the demand of Polish households for the services of creative industries, should be preceded with

[11]They are annually presented by the Central Statistical Office of Poland in the section 'Living Conditions'.

detailed surveys of the above categories of expenditure. Use of the collective entry 'culture and leisure expenditures' may produce misleading conclusions about the potential of the Polish market. Expenditures on purchases of products manufactured by sub-sectors of arts and crafts and creative industries are included mostly in the 2nd, 4th, and 5th categories.

Over the period 2008–2014, average monthly expenditures per capita for recreation and culture fluctuated between 71.86 PLN in 2008 and 70.13 PLN in 2014.[12] The level of per capita expenditures is highly differentiated in terms of regions. For instance, households from the Mazowsze region (which includes Warsaw) spend a monthly average of 114.97 PLN per capita, whereas households from Świętokrzyskie (a peripheral region in East-Central Poland) spend a monthly average of 47.76 PLN per capita. The level of the per capita expenditures depends on the size of the household, location (urban versus rural areas), level of education, and labour status.

The project entitled *Social Diagnosis [Diagnoza społeczna]*, which provides an assessment of changes in the consumer behaviour of Polish households in the creative industries' market, constitutes an important source of data which offers a deeper insight into the issue in question. Studies show that in 2011 between 13% and 20% of the surveyed households (Czapiński and Panek (2011), pp. 113–119) were forced to resign from going to the cinema, theatre, opera, musical theatre, philharmonic (concerts), visiting a museum or an exhibition, or buying a book or newspaper for financial reasons. It is worth stressing, however, that the relatively low percentage of cases of resigning from such activities or purchases for financial reasons was predominantly due to the decreasing interest in these forms of cultural engagement.

If we examine households by type, single parent and multi-child families reported the highest rate of resigning from selected forms of involvement with culture, exceeding even 40%. In 2011, no relevant changes took place compared to the situation 2 years earlier with respect to resigning from participation in culture in the surveyed types of households.

8.2.4 Key Changes in the Sphere of Entrepreneurship

The activities in the sphere of *entrepreneurship* identified during the post-crisis period in the Polish economy are presented in Fig. 8.4.

According to the data of the Global Entrepreneurship Monitor (Polish Agency for Enterprise Development 2014, p. 9), today every fifth adult Pole is planning to start a business within the upcoming 3 years. Though this rate is higher than the EU average (where only 15.9% of adult citizens have the same plans), it is significantly lower than in 2011, when 27% of adult Poles were planning to start their own

[12] 4 PLN is approximately equal to 1 euro.

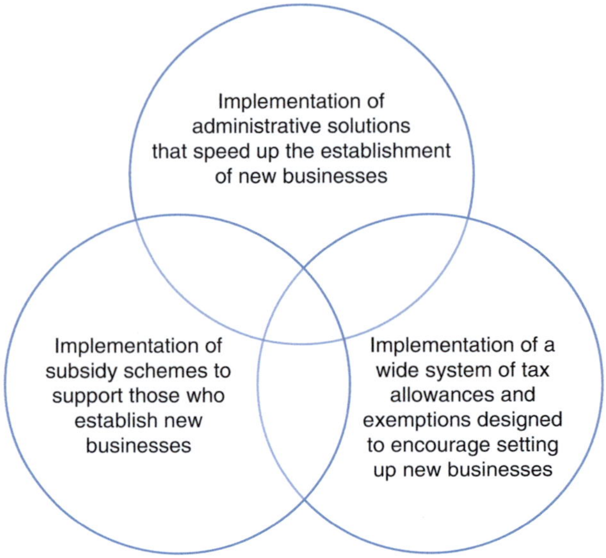

Fig. 8.4 Changes in the sphere of entrepreneurship (Source: Author's studies)

businesses. Fewer people than in 2011 see any chances of a business being started in their immediate environment (a drop from 33% to 26%). Poles highly assess their business knowledge and skills. The majority (52%) claim they are well prepared for this task, while in the EU the same opinion is expressed by slightly more than 42% of adults. Entrepreneurs are noticed in the media by 58% of adult Poles (the EU average is 9% lower). What's more, almost 67% of Poles share the opinion that one's own business is a good career path, while in the EU only 56.9% of adults think along the same lines.

In terms of administrative requirements, starting a business in the creative sector is a relatively easy task. Depending on the business model, there is a multi-stage procedure that must be complied with, but it can also be done online. Obviously, the decision about starting a business in the creative sector is influenced by the operating environment and conditions, which in the post-crisis era include (Kasprzak 2013a, b, p. 2):

- *Education of persons/entities within the creative sector* – most universities (schools) that educate people working in the creative sector do not prepare them to run a business. The efficiency and adequacy of such education is today a crucial issue, closely linked to the discussion about the system of tertiary education in Poland generally. Hence the question whether, for example, fine arts academies should substantively train their students in arts entrepreneurship is a very pertinent one.
- *Competences of persons/entities within the creative sector* – one may often come across criticism about the lack of competences in areas necessary to run a business, such as legal and tax issues, copyright law, fund-raising skills (and not only from public sources), and the ability to plan organisational

development. It is rather unlikely that a single person could accumulate sufficient competence and knowledge that would result in the synergy of creative and business skills, which *de facto* means creative individuals need to seek business partners who could enhance their potential. It should be noted that encouraging persons/entities to enter into lasting business partnerships is desirable in this professional group.

- *Stereotypical perception of creative individuals in the society* – representatives of the creative sector often stress that the effort engaged in creation is frequently not appreciated by society, as the act itself escapes a formal framework and standardisation. When persons performing a 'regular' job compare their work with that of a creative worker, they focus attention on the fact that the working hours of the latter are not regulated; that he/she often dresses in a way that distinguishes him/her from other people; and that they frequently expect gigantic remuneration for drawing just a simple graphic. This latter observation refers to the problem of remuneration for the creative process.[13] This problem touches upon many aspects of entrepreneurship within this social group, such as, *inter alia*: the need to convince the client to pay a 'high' remuneration; the need to convince the client that the preparation of creative content (e.g. an advertisement slogan or a creative act) involves costs and that the potential recipient should pay for the act of creation even if he/she is not going to use it; the need to develop a model to estimate the costs and potential profits of creative work which could be used when seeking to establish working relationships with investors.
- *Educating the client* – people from the creative sector draw attention to the fact that winning over a client is connected with the need to educate her/him. This observation corresponds with other studies confirming that the consumption of cultural services (i.e. products of the creative industries) is a complex process, in which not only household affluence but also the advancement of the cultural education of its members, social status, and the status to which the household aspires, as well as available time and many other factors, figure into the equation.
- *Official collaboration with the creative sector* – representatives of the creative sector often stress that artists' enthusiasm frequently becomes dissipated in formalised procedures involving, *inter alia*, public procurement. Proposals aimed at working out standards for public procurement for specific products from the creative sector, promoting contract award procedures other than the standard ones in use, or expanding the assessment criteria to reduce the role of price as a fundamental factor all seem to be valid as ways of supporting creative potential and motivation.

In light of the above, significant areas connected with the development of an economic policy focused on mobilising entrepreneurship in the creative industries

[13]This phenomenon is discussed in-depth in the literature. See, *inter alia*: Florida (2002), Throsby (2010), Towse (2011). A very interesting account of the subject was also presented by Jung (2011) in the introduction to his publication.

include the following activities undertaken by local authorities and central administration:

- subsidy schemes for start-ups;
- deregulation of some professions;
- tax allowances (exemptions from social insurance premiums) available to start-ups;
- streamlined formalities connected with establishing a business;
- a lower capital threshold for limited liability companies.

8.2.5 *Key Changes in the Sphere of Availability*

The activities supporting the creative sector in the sphere of *availability* are shown in Fig. 8.5.

The access of operators in the creative sector to key resources is determined by their legal status. NGOs and cultural institutions may finance their activities from a wide range of subsidy mechanisms, while the funding of enterprises is subject to the State aid guidelines from the European Union. The sources of funding available to the creative sector are presented in Fig. 8.6.

Mechanisms of *non-returnable funding* include various subsidy schemes. In addition to the already-discussed grants offered by the Minister of Culture and National Heritage as state sponsorship, in the years following the financial crisis, entities in the creative industry sector could benefit from the support of:

- The Polish Agency for Enterprise Development,
- Regional Operational Programmes within the framework of the EU cohesion policy,
- County Labour Offices,
- Commune (local neighbourhood) administrations,
- Agencies for Restructuring and Modernisation of Agriculture,[14]
- NGOs financed by enterprises or government institutions and offering targeted subsidies to institutions operating in the territory of Poland.

The second source of capital for creative enterprises consists of *debt instruments*, including the offers from both commercial banks and financial institutions funded from public resources, the latter of which were developed after the financial crisis mostly as a result of the increasing role of revolving mechanisms involved in the delivery of assistance from the cohesion policy to small- and medium-sized enterprises. The list of major institutions offering debt support in Poland includes, in particular:

[14]Within the framework of the Common Agricultural Policy's funds earmarked for change of the economic structure in rural areas.

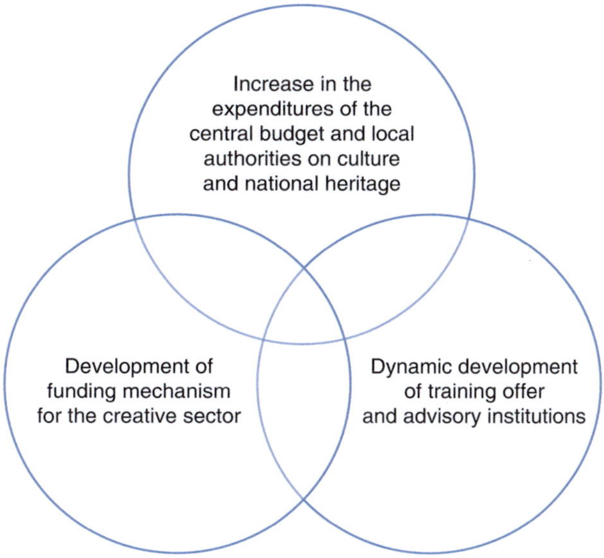

Fig. 8.5 Changes in the sphere of availability (Source: Author's studies)

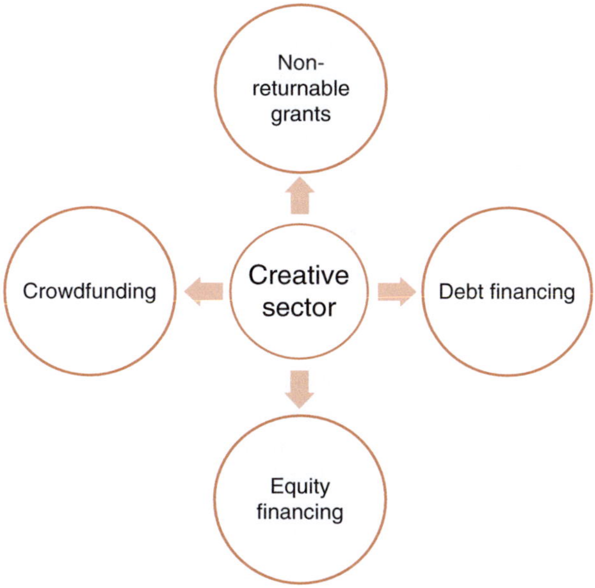

Fig. 8.6 Sources of funding for the creative sector (Source: Author's studies)

- Loan funds, entities granting loans to enterprises from the creative sector as well to as private individuals who run sole proprietorship ventures. Usually each of these funds offer a loan financed using EU funds and granted to enterprises on terms more favourable than those prevailing in the market.
- Guarantee funds, entities giving guarantees for potential borrowers from the creative sector, which improves their creditworthiness and the reliability of the

operator in question when commercial banks assess the risk involved in granting a loan.
- Bank Gospodarstwa Krajowego, which offers loan instruments for small- and medium-sized enterprises, reaching them indirectly via the system of commercial banks.

The third source of funding for the creative sector is *capital financing through seed-capital or venture capital funds*. Such investment funds are financed by both state and cohesion funds (the EU JEREMIE initiative) as well as private investors. However, this part of the capital market in Poland is relatively young and small compared to the developed countries.

The fourth source of funding, increasingly more efficiently used by the creative sector, is *community financing*, which consists of raising money for a particular project using the professional support of a crowdfunding portal, which in turn receives a commission for raising the capital.

From the point of view of developmental needs, the creative sector demands not only access to capital, but access to knowledge as well. Since the outbreak of the financial crisis, institutions which support the growth of creative enterprises and the commercial component of the creative sector have had an important role to play. Those most active within the scope of the creative sector include:

- Local authority organisations, especially the County Labour Offices and Regional Development Agencies
- Economic organisations, especially chambers of commerce and clusters
- Enterprise development centres, in particular training and advisory centres and consultation points as well as incubators of entrepreneurship
- Innovation centres, especially technology transfer centres and technology parks

Many of the above organisations support the creative sector, and specifically the design industry, computer games, and film production, which recently have become quite popular.

The financial offers addressed to the private segment of the creative sector are supplemented with transfers of public resources from the central budget to local authorities. Public expenditures thus come from two sources (Central Statistical Office of Poland 2012, pp. 66–68): the central budget (via Ministry of Culture and National Heritage), and local authorities (via the Voivodeship's Marshal Offices, poviats (local districts) and communes).

In the year 2014 public expenditures on culture and national heritage protection amounted to 9326.8 million PLN – a 13.6% increase compared to 2012 (Central Statistical Office of Poland 2015, p. 95). Of this sum, the state (central) budget expenditures constituted 18.6%, whereas local self-government entities' expenditures constituted 81.4%. This total amount shows that public expenditures on culture and national heritage protection constitute 0.54% of the Gross Domestic Product in Poland. The largest part of these expenditures of the state budget were allocated to the operation of museums (27.0%) and to the protection and preservation of monuments (14.8%). With respect to the budgets of the local self-government bodies, the majority of expenditures were spent on the activities of cultural centres and establishments, clubs, and community centres (29.1%) as well as libraries (16.9%).

8.3 Conclusions

The creative industries are becoming an important part of the Polish economy. Depending on their definition, they can be described as a highly differentiated sector which consists of various types of entities with different orientations. Internal differentiation poses a significant challenge for both national and regional authorities in terms of coordination of their actions towards the sector. The different types of organisations, representing different goals or activities, leads to serious problems with the assigning responsibility to the appropriate administrative body for the design and implementation of an efficient economic policy. For instance, at the national level, one can observe an endless discussion between the Ministry of Culture and National Heritage (as the creative sector generates cultural value) and the Ministry of the Economy (as the creative sector also generates market value) about division of responsibilities and commitments. Similar discussions can be observed at the regional level, where regional authorities are somewhat reluctantly developing inter-departmental task groups oriented towards this sector of activities.

Nevertheless, despite the lack of administrative coordination the creative sector is developing. Its geographical distribution is related to the level of development and capacity of the local economy. As could be expected, the sector prefers well-developed regions with a strong local economy and promising metropolises, thus it is concentrated in the following regions (voivodeships): Wielkopolskie (Poznan), Malopolskie (Krakow), Dolnoslaskie (Wroclaw), Lodzkie (Lodz), and Slaskie (Katowice).

The Mazowieckie region (Warsaw) is an interesting case which requires a more detailed analysis. While it has the most creative sector entities, almost all of them are located in the city of Warsaw. In fact, the city of Warsaw seems to be the main hub for the location of the Polish creative sector, not only due to its implementation of an efficient policy towards it, but also due to the high demand for creative products there and the positive effect of being the capital of Poland.

At the same time, almost all Polish regions now perceive the creative sector as a promising development opportunity with strong growth potential. Statistically however, the size of the sector, calculated on the basis of creative entities registered in the official business register, has not increased faster than the rest of the economy. Creative branches with the highest growth rate are the gaming industry (which has already been identified by the Polish government as a one of the national scientific specialisations), fashion and design, architecture and interior design, advertising and related activities, and radio and music production. The strong growth of those sub-sectors over the analysed period indicates that the creative business sector is becoming more and more interesting and promising for Polish entrepreneurs.

Both the growing size and importance of this sector as well as the recognition of it by policy makers, which has improved over the years, are raising the questions of how to overcome the sector's internal differences and how to design a coherent economic policy towards it?

The IDEA model of economic policy towards the creative sector was developed by the author following various research projects and analyses of the sector. It consists of four policy directions for interventions: (1) strengthening the *internationalisation* of the creative sector, (2) improving and increasing the *demand* for their products, (3) promoting *entrepreneurship* among creative industries, and (4) ensuring the *availability* of the key resources required by them.

In the *internationalisation* dimension, the Polish authorities (both national and regional) are introducing activities aimed at the promotion of the Polish creative sector abroad, stimulating the development of a pro-cluster policy in creative industries, and strengthening the efficiency of those institutions which can assist creative businesses in expanding abroad.

In the *demand* dimension, Polish authorities are developing grant possibilities aimed at the most promising branches of the creative sector, as well as developing patronage-type arrangements and public-private partnerships. These activities are supported by the slow but gradual increase in the nominal expenses of Polish households for cultural and creative products, which however are still below the level observed in developed countries.

In the *entrepreneurship* dimension, Polish authorities are constantly improving a pro-entrepreneurship environment in the Polish economy, with deregulation, reduction of entry barriers, and development of a network of organisations which can help new and creative businesses grow.

In the *availability* dimension, the Polish authorities are increasing the public expenditures on culture and national heritage, developing mechanisms of co-financing for the creative sector, and supporting the growth of training and advisory institutions.

While a longer period is required to evaluate the efficiency of activities mentioned above, nonetheless the Polish creative sector has overcome the world financial crisis quite well and today it seems to have strong foundations for long-term growth.

Appendix

In the NACE rev. 1.2 classification in Poland, the following codes were used to assess the number of a creative sector in the national economy (Table 8.4).

Table 8.4 Creative industries: division and classification by NACE rev. 1.2

Creative industries sector	Sub-sector of arts and crafts	Visual arts (74.20.Z; 90.03.Z; 47.78.Z)
		Performing arts (90.01.Z; 90.02.Z; 90.04.Z)
		National heritage, libraries and archives (91.01.A; 91.01.B; 91.02.Z)
	Sub-sector of creative production	Programming (58.21.Z)
		Publishing (58.11.Z, 58.13.Z; 58.14.Z; 58.19.Z)
		Film and TV production (59.11.Z; 59.13.Z; 59.14.Z)
		Radio and music production (59.20.Z; 60.10.Z; 60.20.Z)
	Sub-sector of creative services	Fashion and design (74.10.Z)
		Advertising and related activities (73.11.Z; 73.12.A; 73.12.B; 73.12.C; 73.12.D)
		Architecture and interior design (71.11.Z)

References

Adamiak P, Dworakowska Z, Herbst J, Przewłocka J (2013) Współpraca w obszarze kultury – samorządy, publiczne instytucje kultury, organizacje pozarządowe [Cooperation in the cultural sphere – local government, public cultural institutions, nongovernmental organisations]. Stowarzyszenie Klon/Jawor, Warsaw

Central Statistical Office of Poland (2012) Kultura w 2011 roku [Culture in 2011]. Warsaw

Central Statistical Office of Poland (2015) Kultura w 2014 roku [Culture in 2014]. Warsaw

Czapiński J, Panek T (eds) (2011) Diagnoza społeczna 2011. Warunki i jakość życia Polaków [Social diagnosis 2011. Conditions and quality of life in Poland]. University of Finance and Management–Social Monitoring Council, Warsaw

Demel K, Falk R, Harauer R, Landsteiner G, Leo H, Ratzenbock V, Schwarz G (2004) An analysis of the economic potential of the creative industries in Vienna. Chamber of Commerce Vienna, Vienna

Department for Culture, Media and Sport (2001) Creative industries mapping document Creative Industries Task Force, London

Dzierżanowski M (ed) (2012) Kierunki i założenia polityki klastrowej w Polsce do 2020 roku, Rekomendacje Grupy roboczej ds. polityki klastrowej [Directions and assumptions of cluster policy in Poland until 2020. Recommendations of the working group for cluster policy]. Polish Agency for Enterprise Development, Warsaw

Fesel B, Sondermann M (2007) Culture and creative industries in Germany. German Commission for UNESCO, Bonn

Florida R (2002) The rise of the creative class: and how it's transforming work, leisure, community and everyday life. Basic Books, New York

Gibbon C (2011) How much can the creative industries contribute to regional development in Britain? Bop Consulting, London

Głowacki J, Hausner J, Jakóbik K, Markiel K, Mituś A, Żabiński M (2009) Finansowanie kultury i zarządzanie instytucjami kultury [Funding culture and managing cultural institutions]. Cracow University of Economics – Malopolska School of Public Administration, Krakow

Grochowski M (2010) Sektor kreatywny w Warszawie. Potencjał i warunki rozwoju [The creative sector in Warsaw. Capacity and conditions for development]. Creative Metropoles, Warsaw

Grochowski M, Mackiewicz M, Michorowska B, Śliwka A (2009) Analiza potrzeb i rozwoju przemysłów kreatywnych [Analysis of the needs and development of creative industries]. Ecorys, Warsaw

Grochowski M, Dudek-Mańkowska S, Fuhrmann M, Zegar T (2012) Sektor kreatywny w województwach pomorskim i kujawsko-pomorskim [The creative sector in the Pomorskie and Kujawsko-Pomorskie regions]. Pomeranian Development Agency, Gdansk
Howkins J (2001) The creative economy: how people make money from ideas. Allen Lane, The Penguin Press, New York
Jung B (ed) (2011) Ekonomika kultury. Od teorii do praktyki [Economics of culture. From theory to practice]. National Centre of Culture, Warsaw
Kasprzak R (2013a) Przemysły kreatywne w Polsce – uwarunkowania i perspektywy [Creative industries in Poland: determinants and prospects]. Kamon Consulting, Warsaw
Kasprzak R (2013b) Raport z warsztatów dla przedstawicieli sektora przemysłu kreatywnego w ramach projektu Creative Poland w dniu 26.03.2013r. Report from the workshop for representatives of creative industries within the Creative Poland Project on 26.03.2013. Creative Poland Project, Warsaw
KEA European Affairs (2006) The economy of culture in Europe, Study prepared for the European Commission (Directorate General for Education and Culture). KEA European Affairs, Turku School of Economics, MKW Wirtschaftsforschung GmbH, Bruxelles
Marshall A (1920) Principles of economics, 8th edn. Macmillan and Co, London
Ministry of the Economy (2014) National scientific specialisations, Warsaw
Musgrave RA (1959) The theory of public finance. McGraw-Hill, New York
Pigou AC (1920) The economics of welfare. Macmillan, London
Polish Agency for Enterprise Development (2014) Global entrepreneurship monitor Polska. Report 2013, Warsaw
Porter ME (2001) Porter o konkurencji [Porter on competitiveness]. PWE, Warsaw
Reis ACF (ed) (2008) Creative economy as a development strategy: a view of developing countries. Itaú Cultural, São Paulo
Samuelson PA (1954) The pure theory of public expenditure. Rev Econ Stat 36:387–389
Senate Department for Urban Development (2008) Creative industries in Berlin. Development and potential. Senate Department for Urban Development, Berlin
Throsby D (2010) Economics and culture. National Centre of Culture, Warsaw
Towse R (2011) A handbook of cultural economics. National Centre of Culture, Warsaw
UN-UNCTAD (2010) Creative economy report 2010: a feasible development option. UN–UNCTAD, New York–Geneva

Chapter 9
The Contribution of the Creative Industries to the Development of the Romanian Economy and Their Resilience to Crisis

Anda Georgiana Becuţ and Andrei Ionuţ Crăciun

9.1 Understanding the Creative Industries in Romania and Their Resilience to Crisis

In order to understand the dynamics of the creative industries and the manner that creativity influences the development of a community or nation, it is important to outline a history of the creative industries concept within the post-socialist space. Although important steps in defining and measuring creativity have been made, there are still many disagreements on how to define creativity and innovation depending on the economic, social, political and cultural contexts considered. First, there are different opinions in the academic environment on using the appropriate concept to define the "new economy": "creative economy" (Howkins 2001) or "knowledge-based economy" (Druker 1970). Second, there are different definitions and methodologies used to analyse the creative industries. The purpose of this chapter is not to compare the pros and cons of one term or another. Still, it is worth mentioning the difficulty to compare and choose the appropriate term and approach from the viewpoint of the Central and East European countries, which is different from the Western countries' viewpoint. The main reason is the different historical backgrounds of these industries within these respective countries, which have followed two different ways of development, the socialist for the former, compared to the capitalist for the latter. Rene Mae (2015), in her paper on the theoretical approach of creative industries in a former socialist country, showed that even the term "industry" has a symbolic significance which is different in different cultural contexts. Although the official term used in European Union documents is "the cultural and creative sectors", in this paper we will use the term of

A.G. Becuţ (✉) • A.I. Crăciun
National Institute for Research and Cultural Training, 57 Barbu Delavrancea Street, Bucharest 011353, Romania
e-mail: anda.becut@culturadata.ro

© Springer International Publishing AG 2017

C. Chapain, T. Stryjakiewicz (eds.), *Creative Industries in Europe*,
DOI 10.1007/978-3-319-56497-5_9

creative industries, in order to highlight those sectors where the production is made on a large scale and sold on the market (resulting in a private property).

In Romania, the creative industries are a new concept, which started being used in official documents in 2009, for example in the *2009–2013 Strategic Plan of the Ministry of Culture, Cults and National Heritage*. During that period, there was an intense debate in the field regarding the definition and role of the creative industries and sectors in Romania. As the results of the Creative Guild Meeting (2009)[1] showed, there is no agreement within the Romanian society regarding the definition of creative industries. Indeed, certain sectors are not considered to have the characteristics of an industry (visual arts, theatre), while certain industries are considered to have too small a creativity content to be included in the cultural and creative sectors (architecture, fashion, IT). In contrast, a quasi-unanimous agreement regarding the attribute of creative industries is expressed in the case of industries with obvious components of production, distribution and mass-consumption, such as publishing, film and music.

In Romania, these disagreements come from a long tradition of considering culture and economy as two separate fields of the social life. As the KEA study "Economy of Culture in Europe" (2006: p. 103) notes, the "cultural and creative sector suffers from stereotypes when it comes to assessing its economic performance. Culture is often perceived as a non-economic activity. Many believe that culture and the economy are two separate worlds and that they have (or that they should have) nothing in common. Those who recognise the economic value of culture tend to perceive the cultural sector as poor in relation to its economic importance". For many Romanian intellectuals, the "art for the sake of art" vision is still the only approach appropriate to consider the contribution of culture to society; therefore they reject any economic discourse about the creative industries. For some, the notion of "cultural industries" evokes dichotomies such as elite versus mass culture, high versus popular culture and fine arts versus commercial entertainment. In this chapter, we agree with Russel Berman (1984) who speaks about a "desublimation" of art and with Scott Lash (1990: p. 157) who characterises post-modern culture through what he calls "eclipse of the aura", adding that "if cultural modernity is to be understood in terms of the separation and even the transcendence or "aura" of aesthetic realism, then *post-modernity* would be a matter of transgression of the boundaries that separate the aesthetic from other cultural practices and from the social itself".

In this context of contested debates on the pros and cons of using the creative industries concept, existing studies carried out on cultural and creative sectors in Romania have used an extended definition of the creative industries, based on the World Intellectual Property Organization model, because the concept of creativity

[1]The Creative Guild Meeting was the first conference organised in Romania on the topic of creative and cultural industries, at the initiative of the Romanian Ministry of Culture. This event gathered experts and practitioners in different fields, such as architecture, advertising, performing arts, visual arts, new media, printing, etc.

is very much related with the concept of intellectual property. The purpose of these studies was to highlight the relation between culture, creativity and economy, especially in a post-modern environment.

The moment when the concept of "creative industries" was first introduced in Romania matches the period when the economic crisis began to show its effects at the international level. Although it seems to be a coincidence, it is possible that this concept appeared on the public policy agenda precisely because of the economic crisis. During this period, policy makers' attention shifted from economic sectors reliant on public funds to those independent from state's support and can be fostered via less interventionist policy models. In his speech entitled *The effects of the economic crisis on culture* at the conference "Culture and the Policies of Change", European Economic and Social Committee Headquarters, in Brussels on 6–7 September 2010, Péter Inkei[2] emphasised the decline of the production, consumption and investments in European creative industries and asked two key questions: "Will the public-private co-operation in the area of creative clusters and incubator centres regain momentum? Will the continuing enthusiasm for the creative sector result in enriching culture – or the opposite, will this deviate political attention and resources from more conventional spheres of culture?"

The effects of the economic crisis were mainly analysed from the viewpoint of the reduction of state support to public cultural institutions, in sectors such as cultural heritage (museums, historical monuments), performing arts (theatres, opera and operetta houses, symphonic orchestras) or public libraries. Decreases in these sectors' cultural production and consumption were obvious and intensely presented and debated in mass media. Available statistics have strengthened the general perception of a decline in culture following the economic crisis at the international level. For example, the Special Eurobarometer 399 showed that "Since 2007, there has been a general decline in participation in most [cultural] activities; the steepest decline was recorded in watching or listening to a cultural programme on the TV or radio, which fell by 6 percentage points from 78 to 72% (while still remaining the most popular form of participation). Least affected were visits to the cinema, which rose by one percentage point, while seeing a ballet, dance performance or an opera remain unchanged. This general decline might be partly an effect of the financial and economic crisis" (2013: p. 5).

The main reason for this decrease in cultural consumption is the financial cost, too high in comparison with household available budget. As the above mentioned report shows: "cost, as measured by "too expensive" responses, is an obstacle for many Europeans, particularly in Eastern European countries (Romania, Bulgaria and Hungary) and in some of the countries the most affected by the economic crisis (Greece, Portugal and Spain)". A possible explanation could be that economic constraints "burden" the cultural domain as a whole from the creation of values to the consumption behaviour. According to Moldoveanu and Ioan Franc (2011: p 26),

[2]Péter Inkei is the Director of the Regional Observatory on Financing Culture in East-Central Europe, The Budapest Observatory.

"the transition in our country and especially the recession/crisis caused the market to restructure and cultural consumption to change in relation to the constraints of the context and personal circumstances".

The results of the Cultural Consumption Barometer 2014 in Romania show (Becuţ and Croitoru 2014) that the economic crisis has affected the level of participation and consumption in all cultural fields; this level only started to increase again after 4 years of continued decrease. These trends in consumption affected the level of the creative industries' market because, if consumption diminishes, society's material/financial resources also diminish (i.e. the contributions to the GDP decrease) and the market targeted by cultural producers must be resized (Moldoveanu and Ioan Franc 2011).

In the following sections, we shall present the analysis of the Romanian creative industries' evolution, at both the macro and micro levels. The first part of the chapter will be dedicated to their performances and trends at the national level; particular attention will be given to the analysis of the three industries contributing the most to the national economy (publishing, advertising and audio-visual). In the second part of the chapter, we shall analyse the manner in which companies in the creative industries were affected by the economic crisis and what measures they took in order to survive or continue to develop.

9.2 Dynamics of the Creative Industries in the Post-socialist Romania

The study, *The Contribution of the Creative Industries to the National Economy,* was an update of the analysis on the copyright-based industries trend carried out for the first time in Romania in 2007, with the help of W.I.P.O. experts.[3] The main objective of the study was to analyse the creative industries on a large scale (macroeconomics) with a special focus on different sub-sectors (publishing, advertising, cinema, etc). The data were collected from the Ministry of Economy (the activity of economic agents) and from the National Institute of Statistics (international commerce). In our study. we used a set of indicators such as contribution to GDP (value added), contribution to national employment, labour productivity, turnover, profitability, foreign trade, etc.

[3]The research was carried out by a team of researchers (Liviu Chelcea, Delia Mucică, Bianca Bălşan, Anda Becuţ and Cătălin Dărăşteanu) from the Centre for Research and Consultancy on Culture as part of the project *The Promotion of the Entrepreneurial Spirit in the Field of Creative Industries,* co-financed by the European Social Fund, Sectoral Operational Programme Human Resources Development 2007–2013. The research used the W.I.P.O. methodology, including four categories of creative or copyright based industries as follows: Core industries, interdependent industries; partial industries and non-dedicated industries, built based on the Romanian correspondents of the NACE.

There are many ways of defining the creative industries and the usage of the term varies among countries. Different models have also been designed to provide a systematic understanding of the structural characteristics of these industries. Each model uses a different basis for classification into "Core" and "peripheral" industries within the creative economy. In our study, we use the W.I.P.O. copyright model, which is based on industries involved directly or indirectly in the creation, manufacture, production, broadcast and distribution of copyrighted works (World Intellectual Property Organization 2003). The focus is thus on intellectual property as the embodiment of the creativity used for making the goods and services included in the classification. A distinction is made between industries that actually produce the intellectual property and those that are necessary to convey the goods and services to the consumer (United Nations Conference on Trade and Development 2008).

According to W.I.P.O. Guide on Surveying the Economic Contribution of the Copyright Industries 2015 Revised Edition, "The Core copyright industries are industries which are wholly engaged in the creation, production and manufacture, performance, broadcasting, communication and exhibition, or distribution and sale of works and other protected subject matter. The interdependent industries are industries which are engaged in the production, manufacture and sale, and renting or leasing of equipment. The partial copyright industries are industries in which a portion of the activities is related to works and other protected subject matter and may involve creation, production and manufacture, performance, broadcasting, communication and exhibition, and distribution and sales. Non-dedicated support industries are those in which a portion of the activities is related to facilitating broadcast communication and the distribution or sale of works and other protected subject matter whose activities have not been included in the core copyright industries". Figure 9.1 presents the concentric circles of the copyright-based industries, used in our study.

The creative industries development during the economic growth period of the Romanian society can be explained by constant investments in creative production and distribution and by a growth of the consumption power and by the diversification of consumption patterns. Despite this development, some difficulties related to the transition from a socialist to a post-socialist market economy have prevented a more dynamic growth of the sectors. Some challenges are common to all sectors of the creative industries, but some are specific problems to each industry (Becuţ and Chelcea 2009).

The contribution of the creative industries to the national GDP almost doubled from 2002 until 2008, until 2009 when the economic crisis generated an absolute decrease of approximately 1.5%. Overall, the Core industries represented the main contribution to the GDP and to creative production. The other types of creative industries only represented together a contribution of 1.23% to the Romanian GDP in 2002 and maintained this level until 2009. In contrast, the contribution of the Core industries to the national GDP had a more accentuated growth and almost tripled from 2002 to 2009; this trend was barely affected by the economic crisis as seen in Table 9.1.

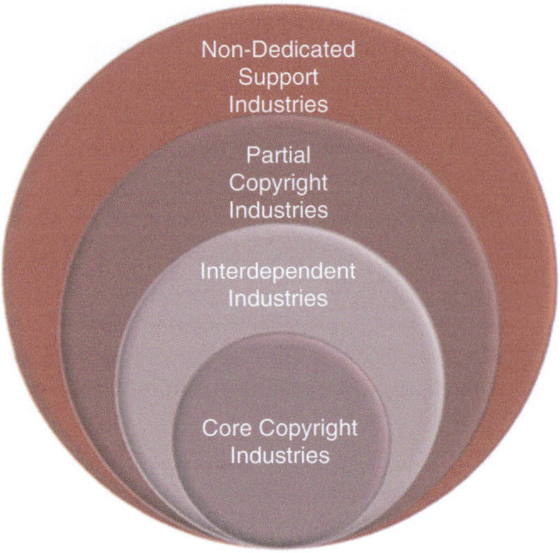

Fig. 9.1 The circles of copyright based industries

Table 9.1 Contribution of creative industries to GDP in the period 2002–2009 (%)

	2002	2003	2004	2005	2006	2007	2008	2009
Creative industries (total)	**3.75**	**5.34**	**4.59**	**5.55**	**6.88**	**6.66**	**7.24**	**5.93**
Core	2.06	3.23	3.33	3.55	5.30	5.21	5.76	4.75
Partial	0.45	0.66	0.53	0.53	0.88	0.79	0.81	0.62
Interdependent	0.96	1.09	0.37	1.08	0.41	0.38	0.40	0.34
Non-dedicated	0.28	0.36	0.36	0.39	0.29	0.28	0.27	0.22

Data source: C.R.C.C. calculations based on data from Public Finance Ministry

Between 2002 and 2005, we identified a substantial increase of 40% in the total number of employees in creative industries – see Table 9.2. After 2 years of decrease (2006–2007), the number of employees in creative industries reached approximately the same level as in 2005, but in 2009 the number decreased by around 20%.

9.2.1 *Labour Productivity*

Creative labour productivity was calculated as a ratio between personnel expenses (thousand Romanian lei) and number of employees and was compared with the overall national productivity. Despite the growth in the number of creative employees, the labour productivity in the creative industries decreased from 2006 until 2009 – see Table 9.3; this cannot be associated with the effects of the economic crisis. It is important to highlight that despite the negative trend (decrease) of the Core and non-dedicated industries, the partial and interdependent industries recorded an increase in labour productivity.

Table 9.2 Contribution to national employment (total number) in the period 2002–2009

	2002	2003	2004	2005	2006	2007	2008	2009
Creative industries (total)	**103.617**	**147.045**	**156.521**	**180.450**	**144.352**	**154.689**	**171.056**	**141.736**
Core	44.019	76.115	89.613	101.801	91.451	99.773	121.299	99.073
Partial	30.044	37.738	36.701	32.257	30.183	31.509	26.877	21.520
Interdependent	17.733	17.203	14.683	24.864	13.940	13.224	14.055	11.129
Non-dedicated	11.821	15.524	16.800	18.528	8.777	10.182	8.825	10.015

Data source: C.R.C.C. calculations based on data from Public Finance Ministry

Table 9.3 Creative labour productivity in the period 2006–2009

	2006		2007		2008		2009	
	Ratio	%	Ratio	%	Ratio	%	Ratio	%
Creative industries (total)	26.770	141	30.292	135	34.067	139	33.479	132
Core	32.573	171	36.636	163	38.199	155	38.356	152
Partial	16.331	86	17.530	78	24.157	98	23.102	91
Interdependent	16.723	88	19.958	89	22.923	93	24.444	97
Non-dedicated	18.165	95	21.035	94	25.200	103	17.561	69
National economy	**19.051**	**100**	**22.425**	**100**	**24.565**	**100**	**25.276**	**100**

Data source: C.R.C.C. calculations based on data from Public Finance Ministry

Table 9.4 Creative industries' turnover dynamics in the period 2006–2009 (2006 = 100)

	2006	2007	2008	2009
Creative industries (total)	100	25.52	25.89	−15.83
Core	100	15.62	20.86	−24.83
Partial	100	3.30	26.05	−22.22
Interdependent	100	20.04	12.86	−17.58
Non-dedicated	100	22.29	24.13	−17.17
National economy	100	18.57	17.34	−14.73

Data source: CRCC calculations based on data from Public Finance Ministry

9.2.2 *Turnover*

The creative turnover fluctuated during the period 2006–2009, with a constant growth in 2007 and 2008 and a decrease in 2009, probably because of the economic crisis – see Table 9.4.

9.2.3 *Profitability*

The profitability of the creative industries presents the same fluctuating dynamics as its turnover and labour productivity over the period 2006–2009; the most accentuated decrease happened in 2009, at the beginning of economic crisis, with a decline of almost 40% in comparison with the previous year (Table 9.5).

9.3 The most Successful Creative Industries in Romania and Their Dynamics

The most successful creative industries in terms of gross value added (GVA) contribution were software, publishing and advertising in 2009. Other relevant industries were film, media, performing arts and music – see Fig. 9.2. The software

Table 9.5 Creative industries' profit share in the total profit at national level in the period 2006–2009 (%)

	2006	2007	2008	2009
Creative industries (total)	6.45	6.18	7.03	4.63
Core	5.1	4.92	5.77	3.84
Partial	0.68	0.63	0.65	0.39
Interdependent	0.37	0.33	0.33	0.22
Non-dedicated	0.3	0.3	0.28	0.18

Data source: CRCC calculations based on data from Public Finance Ministry

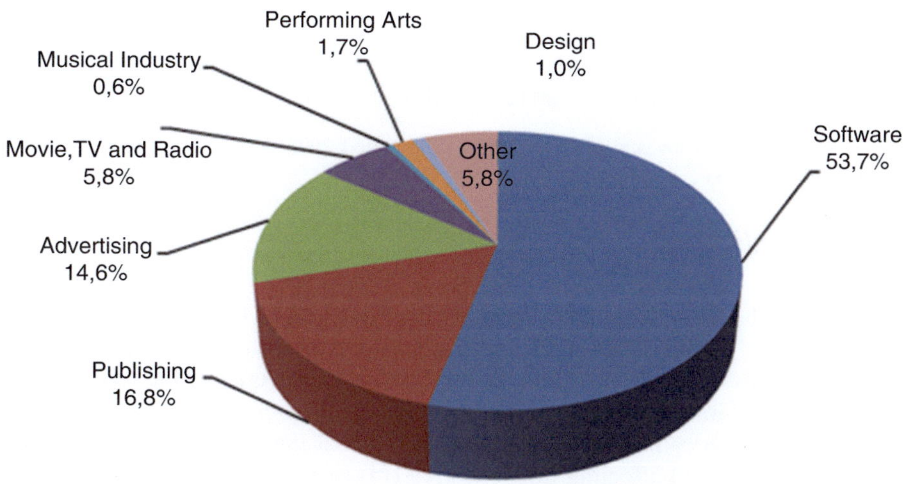

Fig. 9.2 The percentage of gross value added of the creative industries CORE in 2009 (Data source: C.R.C.C. calculations based on data from Public Finance Ministry)

industry displays a constant growth between 2006 and 2008 with the smallest decline from all the industries included in the research in 2009. The sharpest decreases in GVA were experienced in performing arts, advertising, media and publishing.

As mentioned in the first part of this chapter, the overall decrease of the national consumption level in the transition period from a socialist to a market economy resulted in a resizing of the creative industries' market. The level of investments in culture has continuously decreased in the period 2008–2013; this trend seemed to reverse in 2014, with a small increase, similar with the increase of cultural consumption – see Fig. 9.3.

For a better understanding of the trends of the most successful creative industries in Romania, we shall focus on three main industries – publishing, advertising and audio-visual (movie, TV and radio) – over time and highlight the effects of the economic crisis effects registered in the period 2008 onwards.

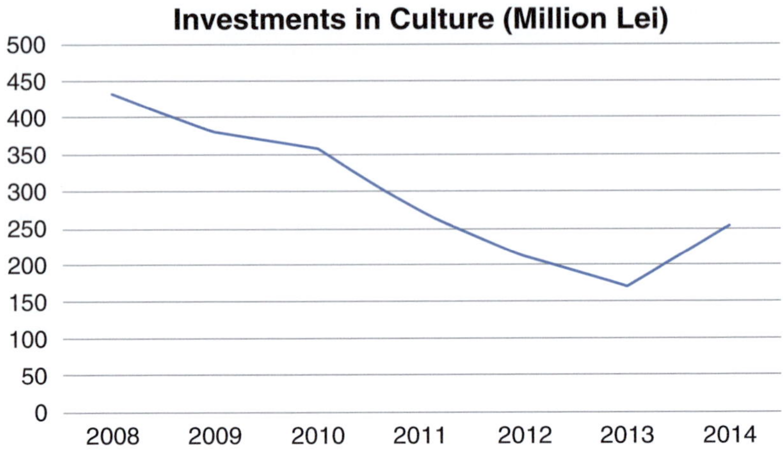

Fig. 9.3 The level of investments in culture in the period 2008–2014 (Data source: National Institute of Statistics)

9.3.1 Publishing

The data available for the publishing industry at the national level show its constant growth in the period 1990–2007 and a small decrease in the first year of the economic crisis – see Fig. 9.4. Though the industry has overtaken the crisis period in 2009–2011, in the last 2 years, there has been a sharp decrease in book production, which can be interpreted by the fact that the crisis effects were slow and that they are still affecting this industry. This explanation is confirmed by the trend in book consumption, expressed in the level of household budget spent on books acquisition. Figure 9.5 shows a small increase of expenses on books, even in the first years of the financial crisis, followed by a slight decrease in the period 2009–2010. Even with a growth in 2014, the average level of the household budget spent on books in a year is very low, around 3 Euro (i.e. 14 lei) or less for a book.

The production of newspapers displays a similar trend as the production of books, with a constant growth until 2011 (except the first year of crisis in 2009) and a sharp decrease in the period 2011–2014 – see Fig. 9.6.

The economic crisis effects are more visible in the trend of the number of employees in Publishing Houses. After a constant growth in the period December 2000–December 2008, in the year 2009, the number of employees decreased under the level of 2007 (around 22,000 people) and maintained this level until December 2012 – see Fig. 9.7. In the last 3 years, we can notice a constant growth, but, with the decrease of production in 2014, this ascending trend might be interrupted.

Because of the lack of official data, foreign trade analysis was carried out only on certain types of creative industries. The data on publishing show a growth of exports even during the period of economic crisis, but a negative balance over the period 2006–2009 – see Table 9.6.

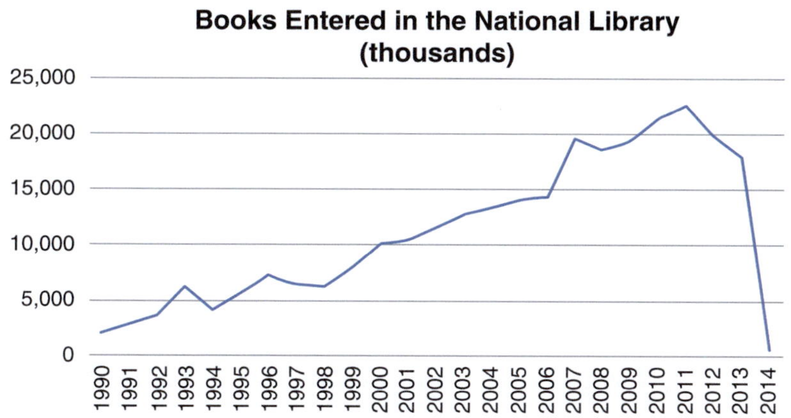

Fig. 9.4 The level of book production in the period 1990–2014 (Data source: National Institute of Statistics)

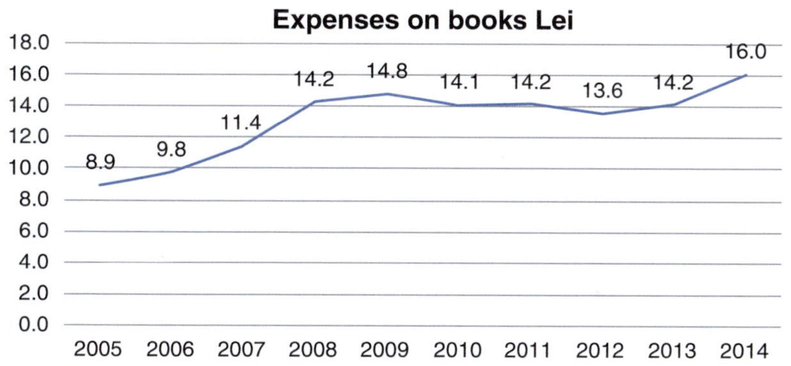

Fig. 9.5 The level of expenses on books in the period 2008–2014 (Data source: National Institute of Statistics)

9.3.2 Advertising

The production of advertising has decreased drastically in the period 2009–2014 (Pîslaru 2015) and this creative sector is in the top of the industries with the sharpest decrease in turnover in this period. Thus, the decrease was not reflected in the number of employees, which raised by 2295 persons from 2009 to 2014 (Idem.). During the economic crisis, the number of advertising hours broadcasted on TV decreased slightly in 2009 and then increased slowly until 2012, when there was a sharp decrease in the number of hours similar to the level registered in 2005 – see Fig. 9.8.

Unlike the trend of advertising distribution on TV, the number advertising hours broadcasted on Radio decreased in the period 2008–2012, as a result of the

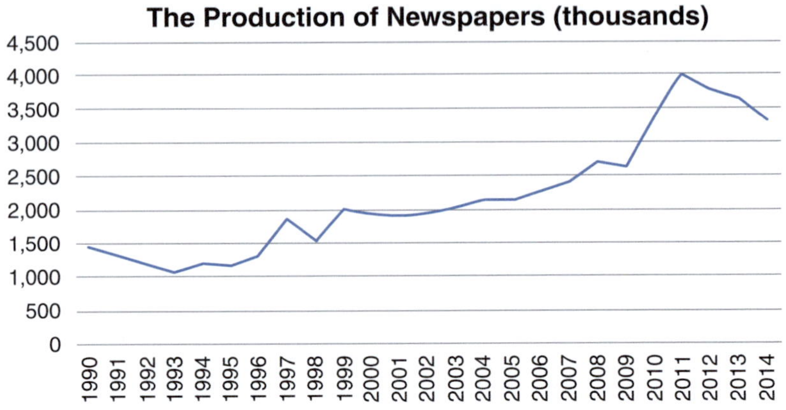

Fig. 9.6 The production of newspapers (thousands) (Data source: National Institute of Statistics)

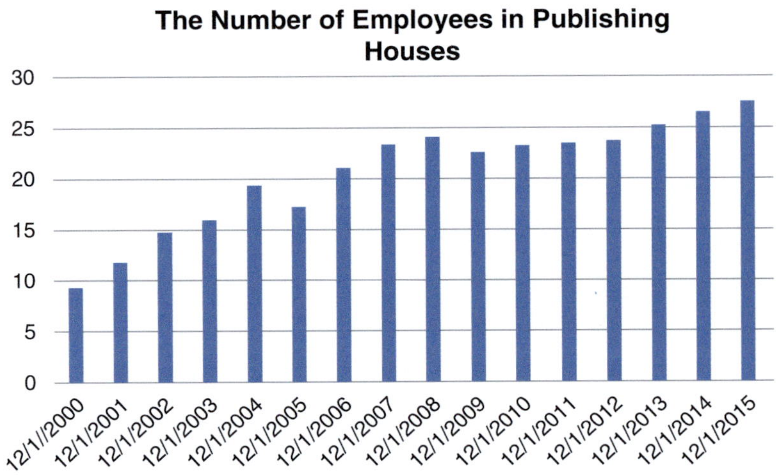

Fig. 9.7 The number of employees in publishing houses (thousands) in the period December 2000–2014 (Data source: National Institute of Statistics)

Table 9.6 Publishing (books, dictionaries) international trade

	Exports ('000 Euro)	Imports ('000 Euro)	Trade balance ('000 Euro)
2006	23.368,84	119.857,14	−96.488,30
2007	27.064,53	116.085,85	−89.021,32
2008	27.215,26	148.392,68	−121.177,42
2009	30.592,23	136.553,79	−105.961,56

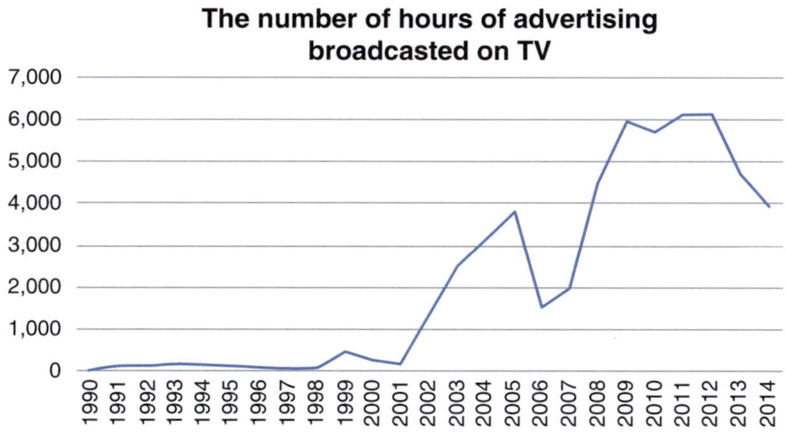

Fig. 9.8 The number of hours of advertising broadcasted on TV in the period 1990–2014 (Data source: National Institute of Statistics)

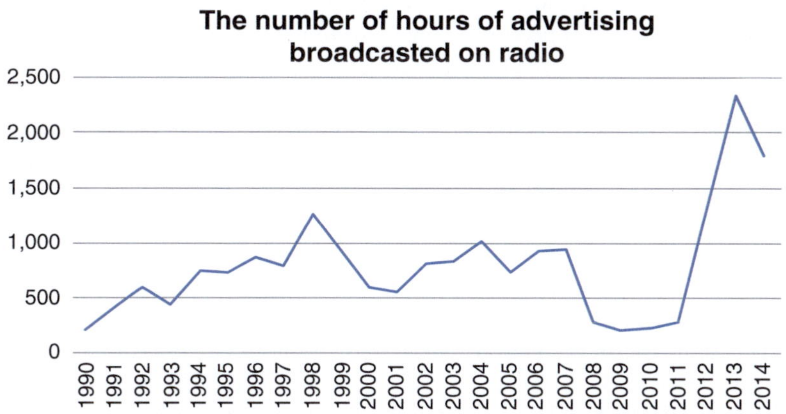

Fig. 9.9 The number of hours of advertising broadcasted on radio in the period 1990–2014 (Data source: National Institute of Statistics)

economic crisis. A sharp growth in the number of hours in 2013 and a decrease in 2014 show that the dynamic of advertising production and distribution is still very fluctuant – see Fig. 9.9.

9.3.3 Audio-Visual (Movie, TV and Radio)

This section presents the dynamics of three sub-sectors: movie, TV and radio. The Romanian movie industry is sometimes considered the best ambassador for the

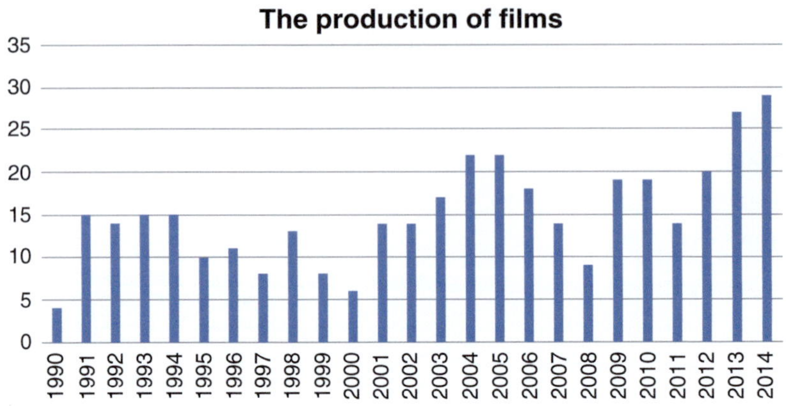

Fig. 9.10 The production of films in the period 1990–2014 (Data source: National Institute of Statistics)

Romanian culture, because of the international prizes won by some Romanian movie production. However, there are critics who consider that the industry is still at a low level and that there are structural impediments for a better growth. The production of movies has been very fluctuant in the last 25 years, with periods of growth (1991–1994, 2000–2006, 2011–2014) and periods of decrease (1994–2000, 2005–2008) with some exceptions – see Fig. 9.10. It is important to note that the highest level of production is still very low compared to other countries with below 30 movies produced in 1 year. These dynamics can be explained by major deficiencies in the movie production infrastructure: few movie sets, old technical equipment (especially sound equipment) and the decrease in the number of technicians and professionals in the industry (Chelcea and Becuț 2010). Another possible explanation for the fluctuant trend in movie production is the low number of spectators in cinemas, due to the constant decrease of the number of cinemas, which was dramatic in the period 1990–2010: from 595 cinemas in 1990 to 67 in 2010 – see Fig. 9.11.

The production in the TV and radio industries increased in the period 1990–2011 despite the economic crisis – see Figs 9.12 and 9.13. However the TV industry lost almost 1500 employees in the period 2009–2014 and 7% of turnover in the same period (Pîslaru 2015).

9.4 Coping with the Financial Crisis (Effects and Strategies)

In this section, we shall focus on the way companies in the cultural and creative industries reacted to the effects of the economic crisis and devised strategies to overcome the difficulties they faced during that period. We shall present the main results of the study "*Effects of the economic crisis on culture*", the first of its kind in

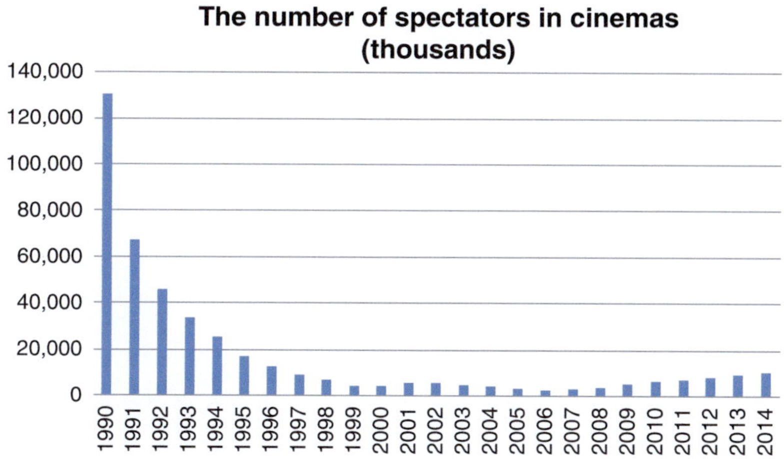

Fig. 9.11 The number of spectators in cinemas in the period 1990–2014 (Data source: National Institute of Statistics)

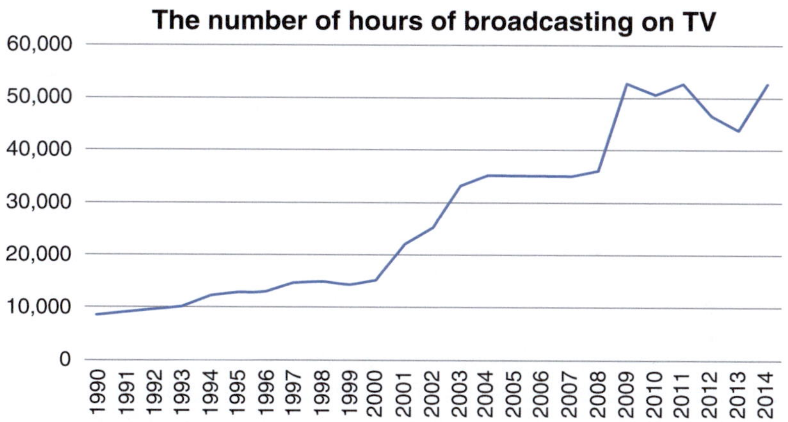

Fig. 9.12 The number of hours of broadcasting on TV in the period 1990–2014 (Data source: National Institute of Statistics)

Romania, conducted by a research team coordinated by the authors.[4] The research objectives were to identify the manner in which the economic crisis has reconfigured the Romanian cultural sector and the strategies to limit these effects, as proposed by the representatives of the companies and organisations that operate in the field. The main research instruments used were: a self-completion questionnaire and semi-structured interviews.

[4]Special thanks to the research team (Bogdan Pălici, Crăița Curteanu, Oana Donose, Ștefania Voicu, Monica Stroe).

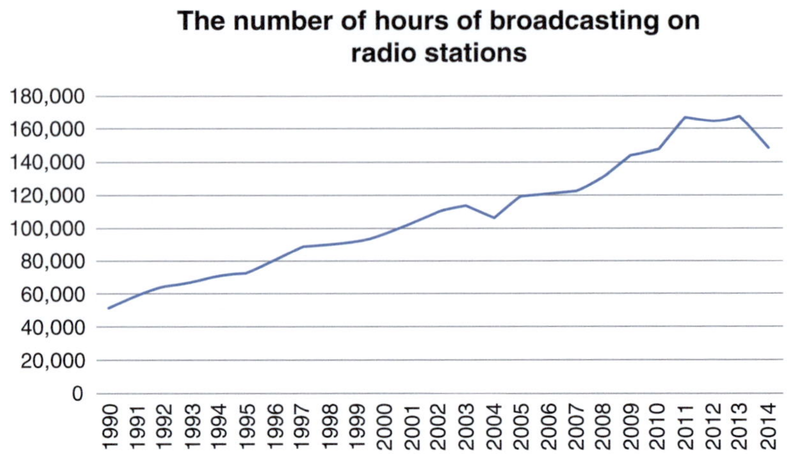

Fig. 9.13 The number of hours of broadcasting on radio stations in the period 1990–2014 (Data source: National Institute of Statistics)

The questionnaire included questions on the changes in companies or organisations' activities as well as in their external environment, detected or predicted by the respondents, on the causes associated with these changes, on the measures they had taken or planned to limit the effects of the crisis and on the institutions in charge with the identification of anti-crisis strategies in the field of culture, in the respondents' opinions. The questionnaire was emailed to 600 organisations that operate in the field of culture: public institutions, non-governmental organisations, collective management bodies and trade associations. The analysis is based on the answers received from 54 of the aforementioned organisations. In addition to the questionnaires, 19 face-to-face, semi-structured interviews were conducted to complement the quantitative data.

The profile of the respondents that filled out the questionnaire reflects the variety of creative activities, their legal status (public, private and non-profit), organisations' sizes (in terms of turnover and number of employees) and of the organisations' income sources. The majority of the respondents[5] (43%) were private companies, followed, in equal measure, by non-governmental organisations and public institutions (both categories with 28%). As regards the cultural activities wherein the organisations operate, the largest amount of information was provided by representatives of the book industry (30%), followed by representatives of museums and performing arts (13%).

[5]The questionnaires were filled out by persons with positions in management or economic departments within the respective organisations. In the report, the category of respondents – where no specifications are made – will be assimilated to the organisations represented by those who filled out the questionnaire.

The activities that recorded a low interest in the participation in the survey were the music and the advertising industries (2% of the respondents each). The majority of survey respondents were small-sized organisations, with a maximum of 50 employees (56% of the respondents). There were also cases (15% of the organisations) where the companies did not have employees, but only collaborators. As far as the financial aspects were concerned, most of the organisations operating in commercial trade (43%) recorded a turnover of less than 500,000 lei (the equivalent of 100,000 Euros), while 17% of the organisations were positioned at the higher limit (more than 10,000,000 lei turnover- the equivalent of 2,500,000 Euros).

The qualitative analysis of this study is based on 19 semi-structured interviews, with organisations across the creative activities four from the music industry field, three from the field of cinema, six from the book industry, two operating in the performing arts and four from the field of visual arts.

9.4.1 Funding Sources

Respondents to the survey highlighted that their funding sources decreased or stagnated, at best for the period January–July 2009. This decrease or stagnation was gradual, though, depending on the possible funding sources used by the cultural actors, i.e. self-financing, public budget, sponsorships, bank loans or fellowships and grants.

In this respect, the decreasing trend was very strong for sponsorship (57%), funding from public budget (48%) and self-financing (47% of the cultural actors considered the incomes coming from this source to be small, as compared to the period January–July 2008). In the sector of written culture, for example, where sponsorships were very important for the set-up of book fairs, the decrease of sponsorships was even more significant: “all the companies we cooperated with in the past years have declined the possibility to cooperate in 2009, for real reasons or using the excuse of the crisis” (representative of a publishers’ association). In contrast, bank loans were the main reliable source of funding with 64% of cultural actors stating their opportunities had remained unchanged and 27% that their opportunities were better compared to 2008. Respondents mostly associated the overall decrease in funding in terms of self-financing (82%) and public budget (80%) to the economic crisis, with some nuances, though. However, the slight decrease in bank loans was mainly regarded as the result of an accumulation of causes pertaining at the same time to the new economic situation and to factors specific to various creative activities, but not as being mainly caused by the economic crisis. This observation complements the idea that bank loans were the income source the least mentioned as being in decline in the cultural sectors.

Respondents who experienced an increase in self-financing and public budget associated this with factors specific to their creative field such as the price of going to a play in the case of theatre:

> ...in the field of theatre the crisis will result in people getting closer to theatre, because I think that the theatre ticket is still affordable and, if you don't have money for other things, you will go to the theatre. The theatre halls are full, at least in Bucharest... (representative of a theatre union)

9.4.2 *Profit*

The evaluations provided by respondents highlight a decrease of both creative turnover (78%) and profitability (70%) for the period January–July 2009, as compared to the same period in 2008. About 83% of the cultural actors considered that this decrease was the consequence of the economic recession. The way the economic crisis and specific elements combined in order to generate a decrease in profit varied depending on the creative sector concerned. The growth in internet access may have been a critical factor. For instance, as far as the music production houses are concerned, even though in the first months of the year, sales are generally lower, their representatives noticed a more pronounced declining trend in 2009 as compared to the previous year. A representative of a music production house told us that niche markets decreased very little during the crisis period compared to the mainstream productions and that the decrease in the sales of Romanian music might be explained by the expansion of the Internet.

> We notice that the sales have decreased as compared to the previous year. The months of February and March are relatively "dead" for us, we expect them to be like that (...). We felt a 10–15% decrease, but only on the mainstream. In the niche market everything is still steady, niches have decreased very little. We started selling more foreign music, the sales of Romanian music have decreased, because of the influence of the Internet – there is a globalisation (...). This has been happening for a while, a few years: because of the easier access to information, the Internet subscriptions are cheap. (representative of a music production house)

9.4.3 *Expenses*

Responses from respondents about changes in their expenses – such as maintenance, rent, providers, product distribution and wages – showed that, unlike the trend for profit and turnover, the level of expenses for January–July 2009 stayed the same as for the period January–July 2008. Thus, the wages (66%), the rent (58%), the distribution expenses (52%), the maintenance expenses (44%) and the providers-related expenses (40%) stagnated. In some sector, like publishing for example, distributor prices were identified as a very important problem.

It is important to take into consideration the impacts of phenomena like depreciation of the national currency or decreases in artistic productions when looking at expenses trends. Indeed, given the changes in the currency exchange rate, the decline in funding sources and the decrease in profit, and considering that many expenses (rent, copyrights, translations, artists' payment, etc.) are made in foreign currency, we may say that the stagnation of the expenses level should be understood, under these circumstances, as a growth of these expenses.

> The expenses have grown in the first months because of the growth of the euro. Being an NGO, I do not have production expenses, but the operating expenses have increased, as the costs have grown. (...). We cannot run away from the crisis, even though the budget is higher than last year, but on the whole the structural expenses will grow more than those in the artistic space (e.g. for projects).(representative of a theatre union).

The majority of respondents considered though that the reasons for the stagnation of their various expenses' level for the period January–July 2009 were most of the time correlated with specific factors of the various creative activities considered. This is the case for distribution expenses (60%), rent expenses (50%), maintenance expenses (50%) and providers-related expenses (40%). On the other hand, only the stagnation of wages was considered a direct consequence of the economic crisis.

9.4.4 *Investments*

In this section, we shall analyse the data related to the evolution or changes detected or estimated by the respondents (cultural actors) for a series of activities commonly performed by the organisations in order to maximise the quality of the cultural services offered to the public/consumers, such as: purchasing specialised technical equipment, purchasing new spaces, investing in human resources professional training and advertising investments.

When comparing the period January–July 2009 to the same period in 2008 in Table 9.7, we notice that all types of investment activities generally recorded a decrease. The highest decreases were recorded for the purchase of technical

Table 9.7 Investments of the cultural organisations: Types of changes that occurred in the period January–July 2009

Investment sector	Type of change occurred in the period January–July 2009		
	Decrease (%)	The same (%)	Increase (%)
Purchase of specialised technical equipment	73	24	2
Purchase of new spaces	52	48	–
Investment in professional training of human resources	66	24	11
Investments in advertising	46	38	16

equipment (73%) and investments in the professional training of human resources (66%). In contrast, the investments in advertising were the least affected (16%).

As regards the identification of the cause of these changes, the interviewed cultural actors mentioned the economic crisis as a decisive factor for the decrease of all activities in the investment sector. The economic context was the cause of the decrease in the purchase of specialised technical equipment for 68% of the respondents. The economic crisis was also mentioned as being the cause for the decrease in investments in the professional training of human resources (76%) and advertising (73%). We notice a different situation for the changes at the level of purchasing new spaces with most respondents attributing this to factors specific to their creative activity (50%) and only a third to the economic crisis.

9.4.5 Human Resources

The collected data show that there was no essential change in human resources, with both the number and the quality of employees and collaborators remaining the same on the whole, for the period January–July in 2008 and 2009. While the economic crisis was the most frequently mentioned as being the cause of the decrease in the number of employees and collaborators, factors specific to each creative activity resulted on the one hand in growing number of employees and collaborators and on the other hand in decreasing trends as regards to quality of human resources. For example, the film sector witnessed important reconversion and brain drain of the sector workforce:

> *Are the collaborators you work with affected, too?* DM: Of course. Most of them have already re-skilled themselves, they left the country. Costume designers, technicians, people who in ten years have received satisfactory training are leaving now, cause they don't find projects, they are going through professional conversion, in order to make money. In France there is a system wherein these people, who only work several months a year, receive money from the state, being encouraged this way to practise these jobs. (film producer)

9.4.6 Partnerships

The number of partnerships initiated locally, nationally and internationally and partner institutions mainly remain *the same* and were overall not really affected by the economic crisis. Three quarter of respondents who mentioned an increase in partnerships in the first half of 2009, as compared to the same period in 2008, explained that this was due to factors specific to their creative activity. For the cultural actors who mentioned a decrease of partnerships at the national level (68%) the cause of this change was the economic crisis. However, these respondents also mentioned an important decrease in the number of international partnerships.

9.4.7 Production of Cultural Goods and Services

Respondents were moderately affected by a decrease in their cultural production for the period January–July 2009. The most significant decrease mentioned was of 55%, and, it was recorded in the production of goods and services, followed by the number of productions/projects (42%). More specifically, 38% noticed a decrease in the number of local productions and 29% for the number of international productions. In addition, quality remained stable for 56% of respondents; only 5% noticed a decrease. The quality-price ratio remained constant for 49% of respondents whereas 18% felt it decreased. After the first two quarters of 2009, experts' evaluations credited the crisis with the prevalent or even exclusive role (we are taking into account a proportion between 75 and 100%) in the decline in cultural goods/services production, in cultural productions/projects, including local and international ones, and in the price-quality ratio.

9.4.8 Distribution of Cultural Products

Most aspects of the cultural products' distribution process remained unchanged in the period January–July 2009, as compared to the similar period in 2008. Around 65% of respondents stated that the number of distribution networks remained the same, 62% considered the quality of the distribution networks to be identical to the previous period, 56% thought the same for local distribution networks and 55% for international distribution network. Those respondents who considered that the distribution aspects were affected by the crisis in the first quarters of the year 2009 in terms of quality and quantity of the distribution network blamed the crisis for this situation and the evolution of international network.

The evaluations were more reserved as regards national distribution networks, as an equal percentage of respondents (44%) asserted that these networks were either less efficient (decreased) or remained unchanged.

9.4.9 Cultural Consumption

In the period January–July 2009, cultural consumption recorded a decrease – as compared to the similar period of 2008 – in the opinion of 60% of the respondents, whereas 40% felt it stagnated. The volume of public/participants also decreased, according to 50% of the respondents and it has stagnated for 32% of them. Some organisations also recorded a growth in the volume of public/participation (18%), but not in the consumption of cultural products and services.

As regards the book industry, the crisis affected the sales in the period considered the most profitable of the year i.e. the Autumn:

> [the sales have decreased since] around November [2008, A/N] as compared to the same period of the previous year. I can even say it started in October. This period, which was the best for sales, started very poorly. (representative of a book publishers and distributors association)

9.5 Conclusion

The concept of creative industries is hard to define and there is a debate on this issue in many countries and Romania is not an exception. Though they were not coined as such, these industries raised and developed in Romania in the communist period and their dynamic reflects the dynamic of the society and economy during the post-socialist period. Their contribution to both economy and society was taken into consideration recently and analyses such as those presented in this chapter are very important because they allow comparisons in time and at the international level (with other countries). This chapter intended to analyse the creative industries in Romania both at the macro and micro levels, in order to highlight the way in which the general dynamics influence the evolution of both the public and private cultural and creative organisations. The results show a fluctuant dynamic of these industries, influenced both by the effects of globalisation and by structural difficulties at the national and local levels affecting creative sectors in different ways. This analysis of the creative industries from the viewpoint of the economic crisis impact is also an important step in providing information to devise appropriate policies and limiting the negative economic and social effects of the crisis. A special attention should be paid to professional training programmes, in order to ensure new skills for professionals who are working with new technologies, both in the process of production and distribution of cultural products. Other policies should facilitate the investments in cultural and creative sectors and the access to finance for small- and medium-sized enterprises, in order to develop new and innovative products. Last, but not the least, public-private partnerships should be encouraged.

References

Becuţ A, Chelcea L (2009) The diagnosis of the cultural sector. Polirom, Bucharest

Becuţ A, Croitoru C (2014) Cultural consumption barometer. Culture, between global and local. Pro Universitaria, Bucharest

Berman R (1984) Modern art and desublimation. Telos 62:31–57

Druker PF (1970) The age of discontinuity. Guidelines to our changing society. Heinemann, London

Howkins J (2001) The creative economy. How people make money from ideas. The Peguin Books, London

Inkei P (2010) The effects of the economic crisis on culture at the conference "Culture and the Policies of Change", EESC Headquarters, Brussels available at https://www.coe.int/t/dg4/cultureheritage/cwe/Effects_Inkei_EN.pdf. Accessed 8 Apr 2016

KEA Affaires Report (2006) Economy of culture in Europe, Available at http://ec.europa.eu/culture/library/studies/cultural-economy_en.pdf. Accessed 20 Aug 2014

Lash S (1990) Sociology of postmodernism. Routledge, London

Ministry of Culture, Cults and National Heritage (2009) The Strategic plan of the ministry of culture, Cults and National Heritage 2009–2013, Bucharest available at http://www.cultura.ro/uploads/files/PLAN_STRATEGIC.pdf. Accessed 14 Oct 2014

Moldoveanu M, Ioan Franc V (2011) The impact of the economic crisis on culture. Rev Gen Manag 4(2):15–21

Pîslaru D (2015) Sectoarele culturale şi creative în România. In: Cojanu V, Croitoru C, Becuţ A (eds) Importanţa economică şi contextul competitiv. Pro Universitaria, Bucureşti, pp 57–87

Rene M (2015) The creative industries: a discourse-theoretical approach. Intl Rev Soc Res 5(2)

W.I.P.O. Guide on Surveying the Economic Contribution of the Copyright Industries (2015) Revised edition http://www.wipo.int/edocs/pubdocs/en/copyright/893/wipo_pub_893.pdf. Accessed 16 Mar 2015

World Intellectual Property Organization (2003) Guide on Surveying the Economic Contribution of the Copyright Industries, World Intellectual Property Organization

Chapter 10
Post-Crisis Spatial Development of Creative Industries: Evidence from Czechia

Ondřej Slach and Jan Ženka

10.1 Introduction

The concept of the creative industries 'arrived' to Czechia approximately a decade ago. The later it arrived, the faster it ranked among other terms often associated with a positive impact on innovation and growth (Cikánek 2009; Kloudová and Chwaszcz 2014). Similar to, for example, the UK (see Campbell 2014), the creative industries are, according to the dominant discourse, an inevitable path to economic prosperity but also could bring other benefits: 'The inclusion of the concept of creative industries in the governmental agenda of Czechia can primarily mean the way out of the land of assembly and trans-shipment points, the country's current label, and can also be an effective remedy for a rather low political and cultural standard, with which Czechia has been dealing with for a long time' (Cikánek 2011, p. 159). However, empirical evidence (Slach et al. 2013) suggests that the nature of creative industries is somewhat different than this 'idealised' concept.

This paper aims to critically evaluate within a wider context the development of creative industries in the post-crisis period between 2009 and 2013. An emphasis is placed, in particular, on changes in the spatial reorganisation of the creative industries in the monitored period. The first part discusses the possible scenarios of the response of creative industries to the crisis and the factors that may affect the probability of accomplishing the discussed scenarios. Subsequently, the data and methods are described. The paper also analyses the impacts of the crisis on the creative industries from an international perspective. Since we have considered the national context to be an important factor in modifying the nature and role of the creative sectors, we pay attention to this in the introductory chapter of the empirical part of the study. The analysis of the development of creative industries in the post-

O. Slach (✉) • J. Ženka
University of Ostrava, Ostrava, Czechia
e-mail: ondrej.slach@osu.cz

© Springer International Publishing AG 2017

C. Chapain, T. Stryjakiewicz (eds.), *Creative Industries in Europe*,
DOI 10.1007/978-3-319-56497-5_10

crisis context is structured around two main lines of inquiry, i.e. sectoral and spatial. The core of the chapter discusses the changes in the distribution of the creative industries at a micro-regional level in Czechia with a particular emphasis on the status of the metropolitan region of the capital city, Prague, in comparison to other metropolitan and non-metropolitan regions. The conclusion discusses the implications of these findings.

In general, we try to ascertain whether the economic crisis has strengthened the position of the largest metropolitan regions by fostering an increase in the spatial concentration of creative industries, or whether the decline in demand in the largest cities and the associated pressure to reduce costs has affected the spatial dispersion of creative industries through the process of suburbanisation.

10.2 Creative Industries and Their Response to the Crisis

Before the economic crisis in 2008, the creative industries were perceived to be one of the new drivers of economic growth in the post-Fordist modes of production and consumption, in line with Scott (2000, p. 3), who notes that 'cultural products of all sorts constitute a constantly increasing share of the output of modern capitalism, and cultural-products sectors represent some of the most dynamic growth industries in the world at the present time'. However, following the financial and economic crisis, questions have been raised with regard to how the creative industries were going to respond to the economic downturn.

10.2.1 Conceptualising the Impact of the Crises on the Creative Industries: Possible Scenarios

With respect to conceptualising the possible impacts of the crisis on the creative industries, Pratt (2009) has identified three possible scenarios.

The first scenario is based on the assumption that the growth or decline of (market-oriented) creative industries depends on the demand from other industries, or from the 'real economy' (p. 495). The drop in the real economy would therefore induce a reduction in private demand and a decline in creative employment and production. Similarly, austerity measures in public funds, which form, in the advanced economies, as much as 50% of GDP, would further reduce the demand and lead the creative industries into a 'vicious circle of decline' (ibid. p. 495). Besides, it cannot be omitted that the decline in demand cannot only be associated with the relationship between the creative industries and the 'real economy', but also with the economic dynamics within the creative industries and within the service sector. Finally, a highly fragmented and unstable segment of the creative industries is dominated by small and micro-businesses, which are highly sensitive to any change in demand (Grodach and Seman 2013). In this first scenario, the creative industries are purely demand-oriented and their ability to respond to the crisis is based on the performance of other sectors.

The second scenario is based on the concept of so called 'creative destruction' (see Schumpeter 1934). In this case, the crisis, on the contrary, would mean an opportunity for the creative industries; due to the creative destruction of existing products and as goods become obsolete, they need to be replaced by new ones. The main driving force of change are creativity-based innovations, which are typical for a group of creative industries. The interconnection of creative industries and 'real economy' can be seen in this scenario as a positive attribute. Companies in the creative industries can be considered as the 'forerunners' in the restructuring of social and economic production forms (Krätke 2002). Due to their low exit/entry barriers and their willingness to take risks, can show higher adaptability to a highly unstable market environment affected by significant changes and can therefore be the bearers of new organisational, social and product innovations.

The essential idea of **the third scenario** lies in the recognition that 'cultural production is driving the development of manufacturing, or the whole economy' (Pratt 2009, p. 496). The main argument is that the economy is increasingly based on symbolic values. In other words, a major part of the value added of products currently consists of their symbolic and aesthetic values and the creative industries have a key role to play in this process (Mossig 2005). The best probable description of the importance of creative industries in this scenario is provided by a design-driven innovation model. In this case, the generation of innovations does not arise as part of intra- or inter-firm research activities, but innovations rather arise within loosely-coupled networks in epistemic (knowledge) communities (of architects, contractors, photographers, critics, curators, publishers, craftsmen, etc.) who give meaning and symbols to specific creative products as well as non-creative products (Verganti 2006). These innovations are dependent on a wider societal and cultural context. Similar importance is paid to the creative industries in the concept of open innovations across sectors, especially in relation to market innovations (Stam et al. 2008; Müller et al. 2009). According to Cooke and De Propris (2011, pp. 373–374), 'cultural and creative industries in this respect can play a crucial role in opening up opportunities to service sectors whilst enabling the upgrading of sustainable high-value added manufacturing activities'. This scenario is also approaching a political interpretation of the significance of the creative industries as a bearer of economic, social and cultural growth in the European Union (EU Commission 2010). Unlike the first scenario, the creative industries are not understood as demand-oriented, but as purely supply-oriented sectors capable of generating new markets through innovation in the wider economy.

The briefly outlined scenarios of the response of creative industries to the crisis are based on a sectoral perspective. However, for their deeper understanding and functionality it is necessary to take into account other factors that may have an influence on the tendency of creative industries to develop according to one of the following outlined scenarios. First of all, this role may be dependent on the specific political context and models of capitalism (variety of capitalism) within which the creative industries operate (Christopherson 2004). Reimer (2009, p. 69) confirms the importance of the political context by stating that 'geographies of creative production (. . .) can be seen to be shaped by regulatory and governance systems at the national level – including the relative responsibilities of labour and employers for skill development and industry sustainability'. The type of economy and the position of the state in the world economic system (and their overall economic performance) may also play a part.

In this respect, for example, a seminal paper by Vernon (1966) can be used as a base. The fundamentals of Vernon's product life cycle theory departs from the assumption that each state or region may have different predispositions to produce a certain product depending on its life cycle. The argumentation of this theory is based on the assumption that the production of a particular product moves to the countries (regions) with the most favourable conditions. In the core countries (e.g. the United States, Germany, the UK), the first phase of the product life cycle is localised – development, creation and launch of the product. Therefore a higher representation of the creative industries with high value added is expected in the core countries. An increasing 'maturity' of the product leads to a decrease in the importance of product innovations and, on the contrary, the importance of process innovations increases, which compared to the development of new products, require considerably less input from the creative sectors. It can be assumed that in semi-peripheral and peripheral countries (with a higher proportion of mature products) there will be a higher probability of an implementation of the first scenario, while in core countries it will rather be the second and the third scenario.

Another possible approach for a more sensitive differentiation of the economic context in which creative industries operate is introduced by the concept of (national) competitiveness. Sala-i-Martin and Artadi (2004, p. 56) identify three stages of competitiveness. A factor-driven stage is characterised by the fact that the primary source of competitiveness for companies is the product price, or low input prices and adopted technology. In the second efficiency-driven stage, the company's productivity results from the quality of its products (no longer just from the product price) and the use of efficient manufacturing processes. The key qualitative characteristic of competitiveness is represented by technological capacity (the company's access to the best available technologies even if they are borrowed from abroad), as well as the efficiency of sub-markets (product, financial and labour markets). In the last innovation-driven stage, thus in the highest stage in terms of quality, the innovation performance, i.e. the ability to create new products and processes with utilisation of the most advanced manufacturing and organisational processes, takes on key importance for the substantiation of a competitive advantage. It can again be expected that the importance and role of creative industries, or the type of scenario will vary depending on the competitiveness stage. In case of the countries, which are in the innovation stage of competitiveness, the role of creative industries as a driver of innovation and economic growth will be the strongest. Therefore, in the crisis period, the third scenario will be the most probable variant, while the first scenario will probably concern countries in the first two stages.

Another aspect of the present debate on creative industries is the spatial distribution and organisation. Creative industries are distributed in space very asymmetrically. Different patterns of response in post-crisis development can be assumed, depending on whether the creative industry companies are located in major cities or in their hinterland, small towns or countryside and whether they are located in spatial proximity of other companies in the creative and other service sectors. The greatest degree of spatial concentration is registered in urban areas, since 'creative industries show an 'urban nature' (Lazzeretti et al. 2008, p. 565). Simultaneously, the proportion of representation of creative industries in urban economies depends on their position in the urban hierarchy with large cities and capitals playing a leading role (Power 2003). However, the position in the urban hierarchy, expressed by the size of the population is not a determining factor. Although there is no consensus whether the creative industries are the cause or, on the contrary, the consequence of growth (Lee 2014), the cities with a higher sectoral specialisation (particularly in manufacturing industries) show the lower representation as well as lower dynamics of the growth of creative industries than the cities that are more diversified (Stam et al. 2008; Slach et al. 2013).

Generally, certain regularity can be traced in the fact that large cities are more diversified, while smaller cities tend to be more specialised (Henderson 1997; Duranton and Puga 2000). Furthermore, smaller cities (or smaller metropolitan regions) have a higher proportion of mature industries, to which specialisation enables better environment for economic growth (Duranton and Puga 2001). Conversely, large metropolitan regions and their urban cores have a higher proportion of new industries and employment is concentrated in the 'high-level service and cultural activities' (Scott 2008: 207). Urbanisation economies are most frequently mentioned as the key rationale for the above-average concentration of creative industries in cities (Lazzeretti et al. 2008; Gong and Hassink 2016). At the same time, large cities offer not only the benefits of urbanisation economies, but also, above all of sub-sectoral specialisations, which are indispensable in terms of the efficiency of incremental innovations (Lorenzen and Frederiksen 2008). The fundamental question is how the spatial characteristics of creative industries play out in their response to the crisis, or possibly to their post-crisis development. The high degree of spatial concentration is an important prerequisite for the reduction of transaction costs (Bathelt and Glückler 2012, p. 229):

(i) **Reducing information costs**. In regions with a high degree of specialisation in a particular industry, the information search and creation costs preceding economic transaction are generally lower. The spatial proximity accelerates the transfer of information affecting local actors and activities. This mechanism is described by Glaeser et al. (2000 p 103) when they state that urban (proximity) density 'facilitates [...] effortless transmissions of ideas and values depend on sight or hearing. Even if the affected person has not seen or heard the influential person himself, it is often true that he knows someone who has had this personal contact. Obviously, the ability to see or hear depreciates sharply with space'. The process of finding suppliers and a workforce has also improved. The highest intensity of transaction costs is typical for the industries with a symbolic knowledge base such as film, music (Scott 2006) and/or advertising (Grabher 2002). These sectors are organised within the so-called 'project ecology' with a high share of tacit knowledge (Grabher 2002), which is characterised by a continual change of team members in the sense 'always change a winning team' (Grabher 2004). This increases transaction costs for both employers and employees (as exemplified by new media, Christopherson 2002).

(ii) **Reducing adaptation costs**. The spatial proximity simplifies the processes of adaptation between firms, which are intertwined in the same production chain. Regular encounters to resolve the arising problems greatly simplify the adaptation to changes. Furthermore, a high concentration while maintaining diversity increases the presence of loose coupling ties that contributes towards (Grabher 1994, pp. 31–34):

   (a) A greater stability of the system, because if there is disruption of part of the system, the negative impacts will only be limited to a given unit without putting the entire system at risk. This reduces the risk that the incident gradually develops into an accident with critical impacts on the entire system.
   (b) A greater sensibility in relation to the external environment, since independent elements of the system can autonomously monitor the external environment, thus contributing to a higher information density and richness. However, too high share of loosely coupled ties within the system can lead to misinterpretations and information chaos (Grabher and Stark 1997).
   (c) The increase in the intensity of the occurrence of variations in the system, arising from the combination of previously disconnected knowledge within the system, i.e. Jacobs externalities (Glaeser et al. 1991), where the importance of generating variations (innovation) increases with an increasing degree of uncertainty (Staber 2001).

(iii) **Reducing communication costs**. Spatial proximity is a key for the transfer of tacit knowledge and trust building, or more generally, for the formation and deepening of 'untraded interdependencies' (Storper 1997). Especially large cities offer a combination of local buzz and global pipelines, while for smaller companies the opportunity to participate in the local buzz does not require any specific investments (Bathelt et al. 2004). Besides, Nachum and Keeble (2003) identified that companies in the creative industries located in the central part of London had a higher intensity of non-local relations than companies located outside the central area.

The above-mentioned spatial aspects of the creative industries suggest that in large cities, which are more diversified and more resilient to the crisis, we can expect a higher probability of the second and third scenarios to the crisis. A strong labour market pool, which increases the degree of adaptability through lower transaction costs, would favour especially the second scenario. Large cities, in this case, act as sort of 'urban incubators'. The probability of the third scenario is supported not only by a higher proportion of new industries and a higher proportion of product innovations and the resulting demand, but also the existence of a greater demand as a whole (public sector, company headquarters, etc.).

The first scenario, on the contrary, appears to be the most relevant for smaller non-metropolitan and metropolitan regions, where the level of relative sectoral specialisation will significantly influence the intensity of the economic decline. However, there are arguments against it. Some smaller metropolitan regions can respond better to the crisis than large metropolitan regions, for example through appropriate sectoral specialisation. In addition, local governments that may increase public expenditures during the crisis, thus stimulating demand, can play a positive role here. Smaller towns also have lower input costs, meaning it might be assumed that there may be a shift of activities into these lower-cost locations. The issue of reduced costs appears highly relevant in terms of de-concentration on an intra-metropolitan level. The cores of metropolitan regions include the so-called diseconomies of agglomeration in the form of high rents, high wages, etc. (Turok 2004). Companies may, in order to reduce their input costs, settle down in the hinterlands of metropolitan regions with lower diseconomies of agglomeration, while they can still take advantage of the urbanisation economies of a core. The benefits resulting from the company's re-location to the hinterland of metropolitan regions are partially derived from the 'borrowed size' concept (see Phelps and Ozawa 2003). The decision of whether the production will be re-located is also the transactional intensity of an industry: in the less transaction-intensive sectors the probability of moving increases and vice versa. Similarly, emerging businesses, especially micro-enterprises can reduce costs by being located in the birthplace of their founder (e.g. see Keeble and Nachum 2002).

Last but not least, the creative industries cannot be considered a homogeneous complex, both in terms of the localisation and the organisation of production (Pratt 2005; Cruz and Teixeira 2015; Tomczak and Stachowiak 2015). That is why the response of individual sectors can be expected to be differentiated. For example, advertising through its focus can respond in advance to a market recovery. Some industries (film, games) may actually grow during the crisis (Pratt and Hutton 2013). In contrast, architecture, for example, may be affected more because it is linked with construction, which is highly pro-cyclical. Reduction of government spending may negatively affect cultural activities funded from public funds or similarly, a reduced consumption of the population may negatively affect book publishing.

10.2.2 Creative Industries and the Crisis: International Empirical Evidence

The response of the creative industries to the crisis has been the subject of some studies which offer a national perspective across Europe. According to TERA

(2014), between 2008 and 2011 employment and value added declined in the creative industries in Spain and the UK. In Spain, the creative industries declined (2009–2012) by 10.73%, which Méndez et al. (2015, p. 11) attribute to the 'freezing consumption, increasing taxes on cultural products and cutting back on governmental grants'. De Propris (2013) offers somewhat different figures in the case of the UK. According to her findings (2009–2010), creative industries have been growing by 2%, while the UK economy has declined. In addition, she points to an increase in self-employed people both inside and outside the creative industries. Indeed, the size structure in the creative industries dominated by micro-firms allows for greater mobility across sectors in comparison to traditional sectors characterised by larger firms. The results of her work would suggest the validity of the second and third scenario discussed previously.

The study by BMWi (2014) suggests that patterns were different in Germany. Overall, while the value added of the creative industries decreased, their employment rate grew (+4.1%). Nevertheless, this varied across sectors. While, for example, advertising or book publishing declined, software/games industry conversely grew by as much as 9% (2009–2012). In contrast, in Italy the value added of the creative industries grew while the employment rate decreased. France reached positive values both in the growth of value added and employment (TERA 2014). The findings of this study imply that examining the response of creative industries to the crisis, the national context or the variety of capitalism in a particular state cannot be ignored. In this sense, Rozentale's work (2014) is of a high relevance for our chapter. The author is studying the impacts of the crisis on creative industries, utilising the example of Riga, the capital of Latvia, even though Latvia differs from Czechia in two aspects. The first aspect is in the degree of international integration, since Czechia represents, according to Drahokoupil and Myant (2015), the so-called FDI-based (second-ranked) type of market economies, while Latvia belongs to a group of peripheral market economies. The course of the crisis also differed since the economy of Latvia declined by an unprecedented 18% between 2008 and 2009, which was almost four times more than the Czech economy (see below in detail). Despite this, Latvia is closer to the Czech context than other Western EU countries mentioned above. Rozentale (2014) found that, on average, the creative industries were more sensitive to the crisis than the entire Latvian's economy, regardless of the fact that Riga is the capital of the country. Although Rozentale's work is a case study, relative contextual proximity suggests that we could expect a similar course in the context of other Central and Eastern European Countries (CEECs), including Czechia, even though the economic downturn there was not as extreme as in the case of Latvia. However, beyond national trends and case study analysis, more detailed regional analysis is also needed to understand the differentiated impacts within countries.

For example, Grodach and Seman (2013) offer an interesting analysis of the American cultural economy in recession, looking at the regional dimension. They identify different patterns in the regional projection of the impacts of the recession on the American cultural economy. Cultural economy hubs such as Los Angeles and New York proved highly resilient, while the intensity of the recession impacts was stronger in smaller metropolitan regions. In these smaller metropolitan regions, 'the region's cultural economy is closely tied to the fate of the major industries there and (. . .) those regions struggling to adapt to larger structural change tend to suffer most' (Grodach and Seman 2013, p. 26). This result would confirm the higher degree of resilience of the creative industries in large cities due to their greater internationalisation and urbanisation economies as discussed in the theoretical part of this chapter. They especially suggest that impacts may be strongly regionally differentiated. At the same time, no evidence was found supporting spatial intra-metropolitan de-concentration of jobs in the cultural economy into the regions offering lower prices.

This very brief discussion of existing research into the creative sector response to the crisis shows that the impacts can be highly diverse. In order to better understand and explain the development of creative industries in the post-crisis period, it is necessary to analyse the context of the studied area. This chapter studied this in detail in the context of Czechia.

10.3 Data and Methods

In this section, we first introduce the methodology we used to define metropolitan and non-metropolitan regions in Czechia. Subsequently, we describe the group of creative industries that we analysed, our data sources and their limits as well as the statistical indicators we used.

Metropolitan regions have been distinguished based on the OECD publication *Redefining Urban Areas (2008)*. This methodology defines metropolitan regions in Europe as a subgroup of urban regions with more than 50,000 inhabitants, while the basic structural unit is LAU2 (Local administrative unit, formerly NUTS 5) municipality. Metropolitan hinterlands were merged from the municipalities with higher than 15% share of employed residents commuting to the nearest core city (OECD 2008).

For our research purposes we employed unique economic data for the years 2009 and 2013, covering the level of municipalities with extended competences (hereinafter referred to as 'micro-regions') – administrative territorial units partially coinciding with local labour areas – were employed for research purposes. They represent the lowest geographic level at which to collect reliable data for employment and value added. Therefore it was not possible to exactly define our (non-) metropolitan regions according to the OECD approach (2008). The borders of the commuting hinterland of metropolitan regions were thus 'aligned' according to the administrative borders of municipalities with extended competences (in line with the approach by Kraft et al. 2014). Non-metropolitan regions were defined as micro-regions in the Czech territory, which do not fall into any of the above metropolitan or urban regions (see Map 10.1).

Due to the superdominant economic position of the Prague metropolitan region in the Czech context (see Ženka et al. 2015), the group of metropolitan regions was

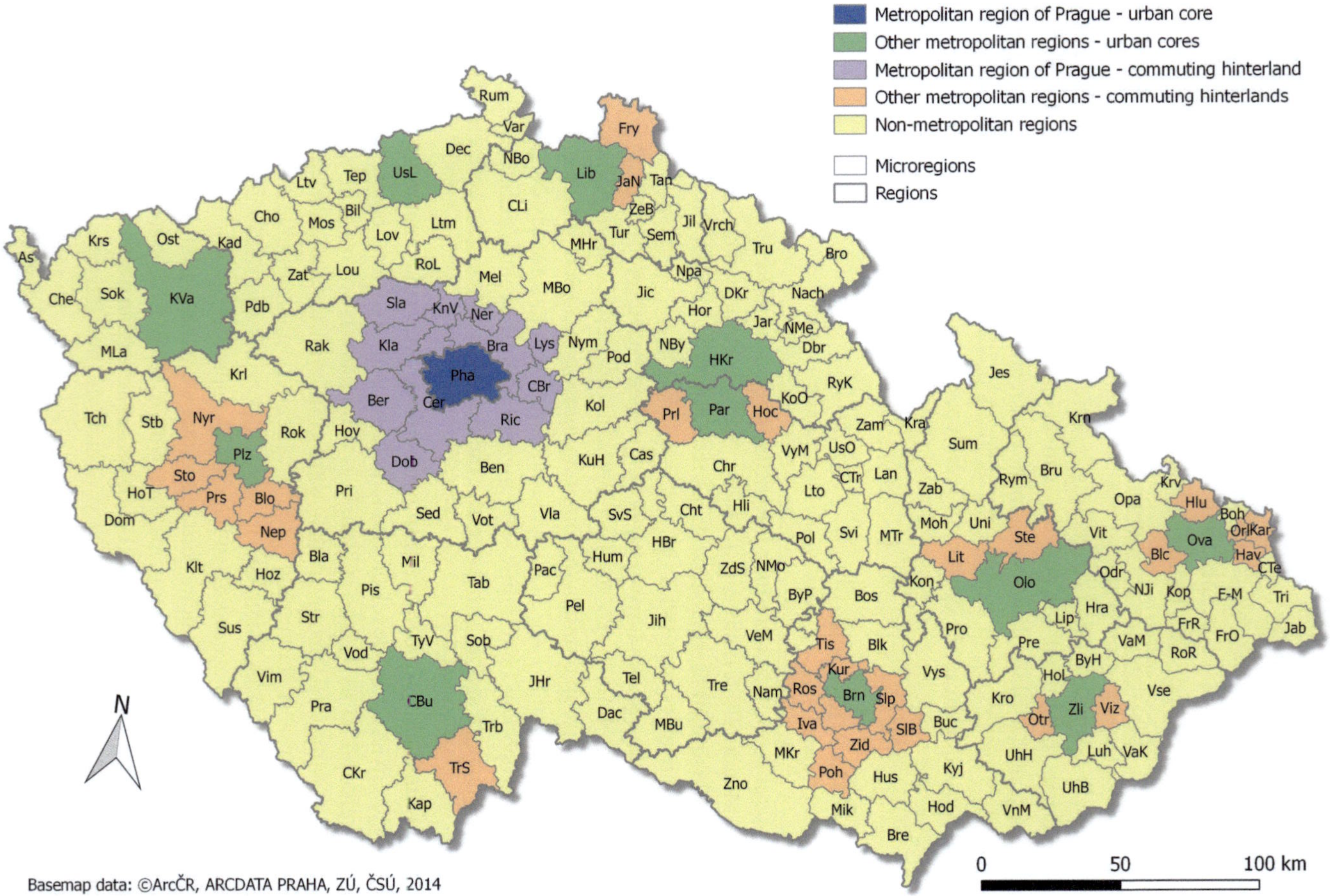

Map 10.1 Delimitation of metropolitan and non-metropolitan regions in Czechia (Source: The authors based on OECD (2008), Hampl (2005) and Kraft et al. (2014))

divided into two subgroups – the metropolitan region of Prague itself and other metropolitan regions. In order to capture the effects of suburbanisation of creative industries, each metropolitan region was further divided into two parts – the urban core and the commuting hinterland. Overall, the Czech territory was divided into five types of micro-regions (Map 10.1):

1. Prague – urban core
2. Prague – commuting hinterland
3. Other metropolitan regions – urban cores
4. Other metropolitan regions – commuting hinterlands
5. Non-metropolitan regions

Change in the distribution of creative industries was evaluated primarily based on the increase or decrease in employment (full-time equivalent) between December 2009 and December 2013. The first date roughly captures the peak of the economic crisis and culminating unemployment (a turn to a lower unemployment rate came in March 2010). The second date demarcates a problematic post-crisis period accompanied by a fluctuating economic performance and unemployment, prelude to the robust recovery of the Czech economy, which came in 2014. The available database did not allow us to analyse in detail the distribution of creative industries over the longer term. Changes in employment rates in the creative industries over a short time period may be difficult to interpret. A decline in employment rate sometimes means a loss of competitiveness of the host region and the transfer of jobs to regions with more favourable conditions for the development of the creative industries. Jobs, however, can also be moved within a metropolitan region – from an urban core to a commuting hinterland with lower prices of land and other advantages – which strengthens a division of metropolitan regions to the core and the hinterland.

A decline of employment, however, can also be caused by a growth in labour productivity and a shift towards creative industries with higher value added – therefore it is not a symptom of the loss, but rather of strengthening competitiveness of the host region and its attractiveness for the development of creative industries. For this reason, we also monitored changes in the value added and labour productivity (value added per employee) in the creative industries between 2009 and 2013.

The primary data source used was the Annual Economic Report of selected production sectors, published by the Czech Statistical Office (CSO 20092013). Data were available for 206 micro-regions and a 2-digital NACE rev. 2.0 sectoral aggregation. The indicators included employment (full-time equivalent) and financial indicators such as production, value added, wages or tangible assets. The database does not cover all sectors of the Czech economy – the data are available for agriculture and forestry, the manufacturing industry, the construction industry and most services (NACE 49–56; 58–64; 66; 68; 69–75; 77–82). Figures for mining and energy, public services and part of the consumer and commercial services (such as retail, wholesale and the repair of motor vehicles) are missing. Total employment in the administrative districts of municipalities with extended powers in the micro-regions is calculated only for the sectors for which there were data available on a micro-regional level; it does not cover the entire spectrum of economic activities. The shares of creative sectors in the regional employment rate must therefore be understood as shares in employment in the above sectors, not as shares of total employment.

Based on the data available, we analysed a group of market-oriented creative industries:

- Printing and reproduction of recorded media (NACE 18), i.e. printing
- Publishing activities (NACE 58), i.e. publishing
- Motion picture, video and television programme production (NACE 59), i.e. film, TV and video
- Programme and broadcasting activities (60), i.e. radio
- Architectural and engineering activities, technical testing and analyses (NACE 71), i.e. architecture and engineering
- Advertising and market research (NACE 73), i.e. advertising
- Other professional, scientific and technical activities (NACE 74), i.e. other activities

Classical creative industries with a symbolic knowledge base prevail in this group (printing, publishing, film, TV and video, advertising). A problematic result of the gross sectoral aggregation is the inclusion of certain service sectors, which have a prevailing synthetic knowledge base and whose characteristics do not quite fit in the creative industries: architecture and engineering (besides architecture itself) and other activities. Printing was also added, although it is a sector with a synthetic knowledge base, however, it belongs to the so-called semi-creative industries (Boix 2013).

Changes in the spatial distribution of creative industries were evaluated by the Herfindahl-Hirschman index (HHI), which was calculated from the relative shares of micro-regions in total employment of each particular creative industry. Formally, HHI is denoted as:

$$HHI = \sum\nolimits_k e_k^2 \tag{9.1}$$

where e_k is the relative share of employment in a micro-region k in total national employment in the particular creative industry or the whole group of creative industries. High values of HHI thus indicate high levels of spatial concentration, while low values signify spatially dispersed industries. Correspondingly, falling values of HHI between 2009 and 2013 document the process of spatial dispersion, while an increasing HHI signifies the process of spatial concentration.

10.4 Creative Industries in Czechia

In the 'varieties of capitalism' concept (Hall and Soskice 2001) Nölke and Vliegenthart (2009) identify a third specific variety of capitalism for Central and Eastern European countries: the so-called dependent market economy. These economies have a comparative advantage in the production of relatively complex durable products, where the source of such a comparative advantage is the institutional complementarities between:

(a) The educated yet cheap labour
(b) The transfer of technological innovation within trans-national corporations (TNCs)
(c) The capital inflow through direct foreign investments (Nölke and Vliegenthart 2009, p. 679)

An important role in this variety of capitalism is played by TNCs, which control the dominant part of the key export-oriented production sectors (automotive, mechanical engineering, electronics), based on an intensive use of the labour force. The preference of neoliberal approaches ('embedded neoliberalism', cf. Bohle and Greskovits 2007) also reinforces the position and role of TNCs. Novotný et al. (2016, p. 16), in their recent work, confirm the validity of a dependent market type of economy, because 'evidence provided on productivity differentials suggest that Czech manufacturing – even though intensively integrated into the global economy – has adopted a specific and semi-peripheral position in the European economy based on the low cost of relatively well-trained labour'. As such, the Czech economy can be evaluated as a semi-peripheral economy with a higher proportion of mature industries, driven primarily by efficiency. Such economies are characterised by a predominance of mostly large series of relatively standardised products (economies of scale) with lower value added and low design intensity (Slach et al. 2013).

To better understand the Czech context, the spatial aspects of economic development cannot be omitted. The transformation process has dramatically transformed the economic landscape of Czechia, which until 1989, was characterised by minimal socio-economic differences (equalisation, at the expense of external competitiveness (Blažek and Uhlíř 2002). Its main in terms of spatial (vertical) differences is the growth of Prague and its hinterlands, similar to all CEECs countries, 'where capital cities capture a large part of the benefits of the incorporation in the global economy by becoming the privileged relays of an economic renewal controlled by the Western firms' (David et al. 2013, p. 246). In terms of the horizontal dimension, worth mentioning is a favourable geographic position close to Western markets with good transport connections (Musil 1993; Blažek and Csank 2007; Petrakos 2001).

Generally, the importance of metropolitan regions shows a growing tendency in Czechia (Čermák et al. 2009; Maier and Franke 2015), which is expressed by their economic growth and/or their development potential (Novák and Netrdová 2011). It is virtually the delayed process of 'metropolisation' in terms of spatial concentration of knowledge-intensive services and creative industries (see Krätke 2007). Ženka et al. (2015) empirically confirm this trend in the spatial distribution of the manufacturing industry, which is significantly represented in the non-metropolitan regions and knowledge-intensive business services (KIBS), which are concentrated in the largest metropolitan regions. The Prague capital itself concentrated in 2009 more than 40% of national employment in KIBS, i.e. more than all the other metropolitan regions together.

Simultaneously, there is an increasing intensity of peripherisation/ marginalisation of some non-metropolitan regions (expressed by an above-average unemployment rate, an aging population, out-migration and poverty), not only those located in the outskirts, but recently also in the inner peripheries (Musil and Müller 2008). This marginalisation trend is supported by a neoliberal regional development policy (deregulation) and the concentration of direct foreign investment in strong rather than weak regions (Drahokoupil 2007; Hlaváček and Koutský 2011). It should be emphasised that regional divergence between the non-metropolitan and metropolitan regions has not been significantly growing since 2000 and conversely, whereas divergence between metropolitan regions has increased. This trend is affected by the existence of two large old industrial metropolitan regions (Ostrava and Ústí nad Labem), which with their size and weak economic performance reduce the overall growth in other metropolitan regions (Blažek and Csank 2007). To sum up, the transformation process in Czechia has brought an increase in regional inequalities (Heidenreich 2003).

Since the creative industries are attracted to growing regions, their distribution to some extent follows the trends outlined above. Empirical studies within CEECs point out a dominant concentration of creative industries in capital cities (e.g. Rehák et al. 2013). An analysis by Slach et al. (2013) identified, based on data from 2009, that in Czechia creative industries are strongly spatially concentrated, primarily in large cities (see Map 10.2). Prague, where almost 40% of all employment of all studied creative industries in Czechia is concentrated, occupies a totally dominant position. Besides, Prague is one of the cities with the highest proportion of creative industries in the labour market throughout the European Union (Power and Nielsen 2010). These findings confirm Prague's position as a so-called global gamma city within world city network (Beaverstock et al. 1999). Prague's high position coincides with a low position of two other largest centres – Brno and Ostrava, concentrating respectively only 9% and 3% of all jobs in the creative industries. The order of these cities is largely determined by the size, geographic location and the position of the particular centre in a settlement and regional hierarchy (Blažek and Csank 2007). Despite the different context, urbanisation economies have a positive impact on the presence of creative industries in the Czech regions, especially if they are combined with the diversified specialisation.

For abbreviations and full names of included microregions see Ženka et al. (2015).

If the spatial organisation of the creative industries and the principles affecting it can, to some extent, be seen as universal, the sectoral characteristics of the creative industries in the post-socialist context may appear to be far from universality. Rozentale and Lavanga (2014), for example, point out that the creative industries in post-socialist countries differ by a lower level of internationalisation, a lower level of demand uncertainty and generally a lower value-added production.

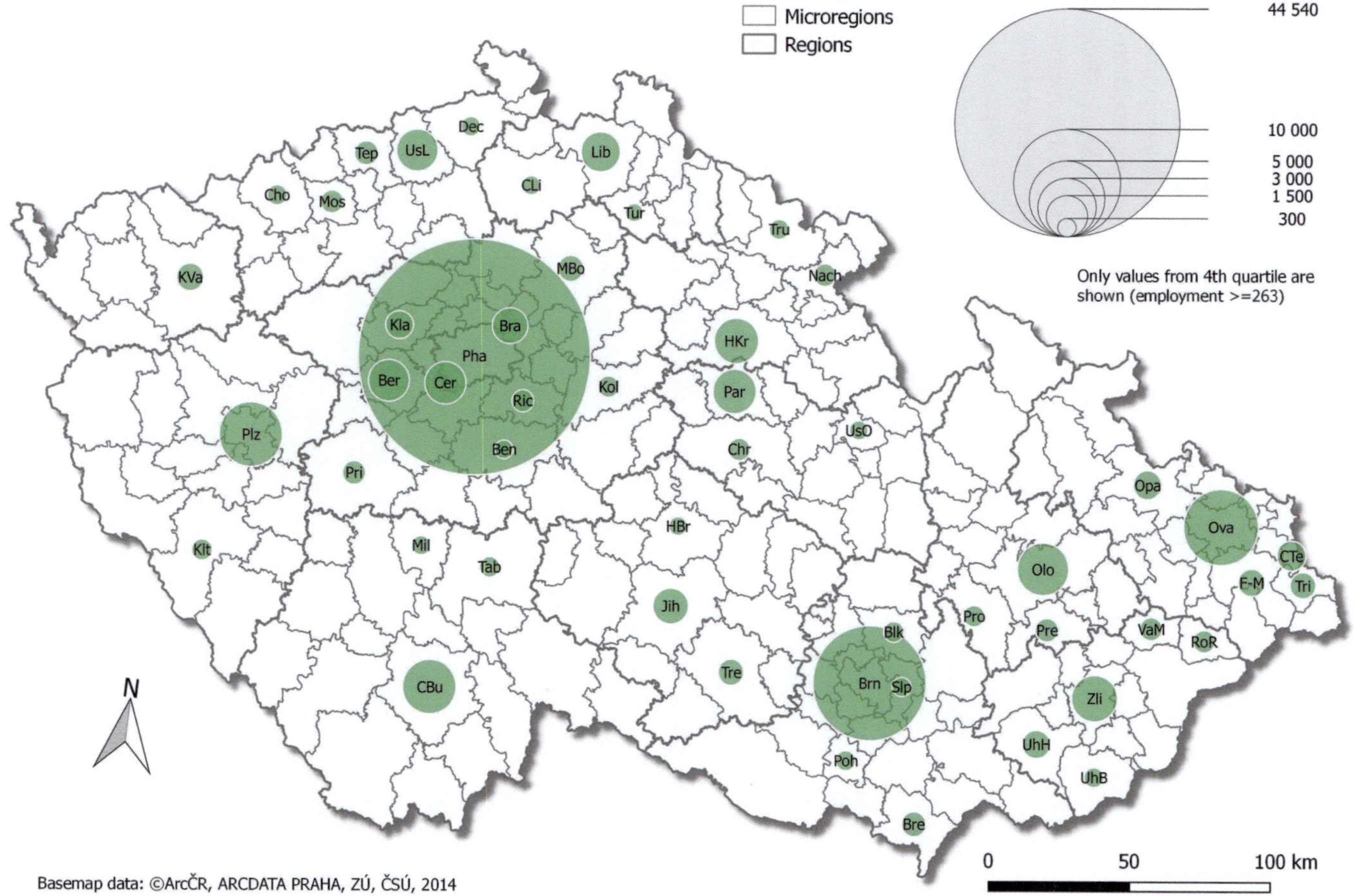

Map 10.2 Spatial distribution of employment at the micro-regional level (2013) (Source: CSO 2013; the authors)

Rumpel et al. (2010) confirm the low level of internationalisation and the high external dependence of creative industries on regional markets.

A unique database of the Czech Statistical Office (2010) provides very interesting information about selected creative industries. In 2010, one of the largest creative sectors in Czechia was advertising. Information about wages (at a company level) shows that in Prague the average wage in this sector (38,689 CZK)[1] exceeded not only the national average (23,231 CZK), but also the regional average (30,100 CZK). In contrast, the average wage in the advertising sector in non-metropolitan regions (18,800 CZK) was significantly lower than the national average. Additionally, more than 88% of employees working in this sector in non-metropolitan regions were below the wage threshold, while in Prague only 44.1%. Other metropolitan regions did a little better, but even there the average wage in advertising (21,371 CZK) was below the national average. Large differences were also observed in the level of production per employee. In non-metropolitan regions it only reached 51% of the Prague level; in other metropolitan regions, this proportion was even as low as 40.3%. So, it seems that the advertising sector creates, especially in non-metropolitan regions, poorly paid jobs with low productivity. The next question is, how the crisis has affected these differences.

10.4.1 Post-Crisis Development of the Czech Economy and the Creative Industries at the National Level

The world economic crisis was reflected in Czechia by an annual decline in GDP of 4.7%, which is less than in Hungary, but more than in neighbouring Poland (see Myant and Drahokoupil 2012). Between 2009 and 2013, the Czech GDP growth circumscribed a W curve. Besides the strong pro-export orientation of the Czech economy, the negative GDP growth was also influenced by an austerity policy of the right-wing government (2010–2013). However, since 2014 a strong economic recovery is observed. The question is how the creative industries performed in the post-crisis period in comparison to other sectors such as manufacturing, construction and knowledge-intensive business services (KIBS).

While the employment rate in the manufacturing industry has remained virtually unchanged, the KIBS sector, as much more important customer for creative products and services, grew by 14% between 2009 and 2013 (Table 10.1).

Nevertheless, the actual creative industries registered an overall decline in the employment rate by 7% in the post-crisis period (Table 10.1). The most poorly performing industry in the post-crisis period was the motion picture and video industry, followed by the printing and publishing industries, radio and TV. The former declined due to the tightening in public budgets (grants, public television

[1] 1 CZK is approximately equal to 0.04 euro.

Table 10.1 Employment change at industry level 2009–2013

NACE sectors	Employment		Dynamics index (2009 = 100)
	2009	2013	
Manufacturing	1,066,195	1,051,174	99
High-tech manufacturing	50,695	48,854	96
Construction	260,501	218,338	84
KIBS[a]	248,333	283,150	114
Creative industries (CI)	**116,558**	**108,412**	**93**
Printing	19,396	16,543	85
Publishing	14,253	12,103	85
Film, TV and video	2044	1497	73
Radio	1553	1313	85
Architecture and engineering	49,855	48,235	97
Advertising	18,526	15,805	85
Other activities	10,930	12,916	118
All industries	2,179,124	2,108,790	97

[a]KIBS – Knowledge-intensive business services
Source: CSO 2009, 2013

production) as well as to a limited amount of incentives to foreign filmmakers. Decrease of the latter can be attributed probably to the decline in demand.

Low decrease of unemployment in the NACE 71 'Architectural and engineering activities' can be attributed to the growth of engineering activities, which serve primarily the manufacturing firms. Engineering activities do not fall into the narrowly defined creative sector, but their share in the total NACE 71 employment is roughly 85% (NIPOS 2011).There is a reason to believe that the slight decline in the NACE 71 employment rate was caused by falling employment rate in architectural services. Their main customer – construction sector – exhibited also a significant decrease in the number of jobs.

The reason for employment growth in other activities can be attributed to increasing demand for translation and interpretation services, forming the major part of the NACE 74 industry, while design industry accounts for only ca 600 jobs (NIPOS 2013).

10.4.2 Micro-Regional Perspective of Post-Crisis Development

The core of this chapter is the analysis and evaluation of the spatial dynamics of creative industries in the post-crisis period. In the first spatial dimension we have studied whether there was an increase or a decrease in the rate of geographical concentration of creative industries. We have found that spatial concentration of both employment and value added increased in the period 2009–2013.

Table 10.2 Spatial concentration of employment in the creative industries (Herfindahl-Hirschman index – HHI)

NACE sectors	2009	2013	HHI (2009 = 100)
Creative industries (CI)	**1545**	**2390**	**155**
Printing	1763	1843	105
Publishing	951	871	91
Film, TV and video	3372	3714	110
Radio	4671	5583	120
Architecture and engineering	7065	7316	104
Advertising	1263	1279	101
Other activities	3337	3507	105
All industries	434	502	116

Source: CSO 2009, 2013

We have identified two basic groups of the creative industries according to their spatial development in the post-crisis period. In the first group of relatively evenly spatially dispersed industries (see Slach et al. 2013) the rate of spatial concentration decreased (publishing) or only slightly increased (architecture, advertising, other activities). In the second group of symbolically oriented industries, the rate of spatial concentration grew, indicating the decline of employment outside Prague. Spatial (de)concentration of value added in creative industries literally corresponds to the trends in employment. Despite spatial dispersion of publishing, the group of all creative industries together experienced a rapid increase in spatial concentration (Table 10.2).

The second spatial dimension focuses on the development of creative industries at the (micro)regional level (Tables 10.3 and 10.4; Map 10.3). We assumed that the share of the creative industries would be declining in the non-metropolitan regions. The findings, however, do not fully confirm this assumption. The creative industries deviated by only 1% from the nation-wide employment decline in non-metropolitan regions. The relatively low intensity of decline can be explained by spatial dispersion of publishing and only modest increase in spatial concentration of printing, advertising and other activities. Not only in non-metropolitan regions but also in absolute figures, the architecture and engineering sector was developing quite well.

Among the reasons explaining the resistance of the above-mentioned creative industries to the crisis in the non-metropolitan regions, one should mention growth in the employment rate in high-tech manufacturing, which forms a significant part of the demand for services in these industries. Again, it needs to be emphasised that the dominant part of the employment in this sector is bound to a technology-related knowledge (synthetic knowledge base) and less to a symbolic content. Conversely, in accordance with the assumptions, radio, film, TV and video in these regions decreased most significantly.

Creative industries in metropolitan regions were falling behind more rapidly than the rest of the economy. The most visible change affected advertising, or film, TV and video, while the only growing industry was NACE 71 – other activities.

Table 10.3 Change in the spatial distribution of jobs in the creative industries 2009–2013

Employment in:	Metropolitan regions				Non-metropolitan regions	Czechia
	Prague – core	Prague – hinterland	Others – core	Others – hinterland		
Manufacturing	12,248	–5425	–6775	3496	–18,566	–15,021
High–tech manufacturing	–763	–3870	–2581	297	5076	–1841
Construction	–3775	–1711	–14,613	–1516	–20,549	–42,163
KIBS[a]	21,835	–459	11,605	72	1764	34,818
Creative industries (CI)	**–2132**	**371**	**–3629**	**–515**	**–2241**	**–8145**
Printing	–1174	359	–999	–234	–806	–2853
Publishing	–868	65	–466	–36	–845	–2150
Film, TV and video	–273	–46	–122	–23	–83	–547
Radio	–182	9	–51	1	–17	–240
Architecture and engineering	–480	–131	–530	–197	–282	–1620
Advertising	–1252	207	–1363	–53	–259	–2721
Other activities	2098	–92	–97	26	52	1986
All industries	27,582	–6143	–32,669	–1365	–57,740	–70,334

Source: CSO 2009, 2013
[a]KIBS – Knowledge-intensive business services

Table 10.4 Change in the spatial distribution of value added in the creative industries 2009–2013 (million CZK)

	Metropolitan regions				Non-metropolitan regions	Czechia
	Prague-core	Prague-hinterland	Others-core	Others-hinterland		
Creative industries (CI)	1228	579	–2602	–746	–1318	–2860
Printing	–536	–167	–687	–94	–344	–1828
Publishing	2456	824	–51	–7	–409	2813
Film, TV and video	–43	–87	–24	–12	–63	–229
Radio	–1393	–9	–68	–1	–51	–1522
Architecture and engineering	–2518	–190	–1130	–844	–809	–5491
Advertising	–635	66	–425	–37	40	–991
Other activities	3898	142	–217	247	318	4388

Source: CSO 2009, 2013

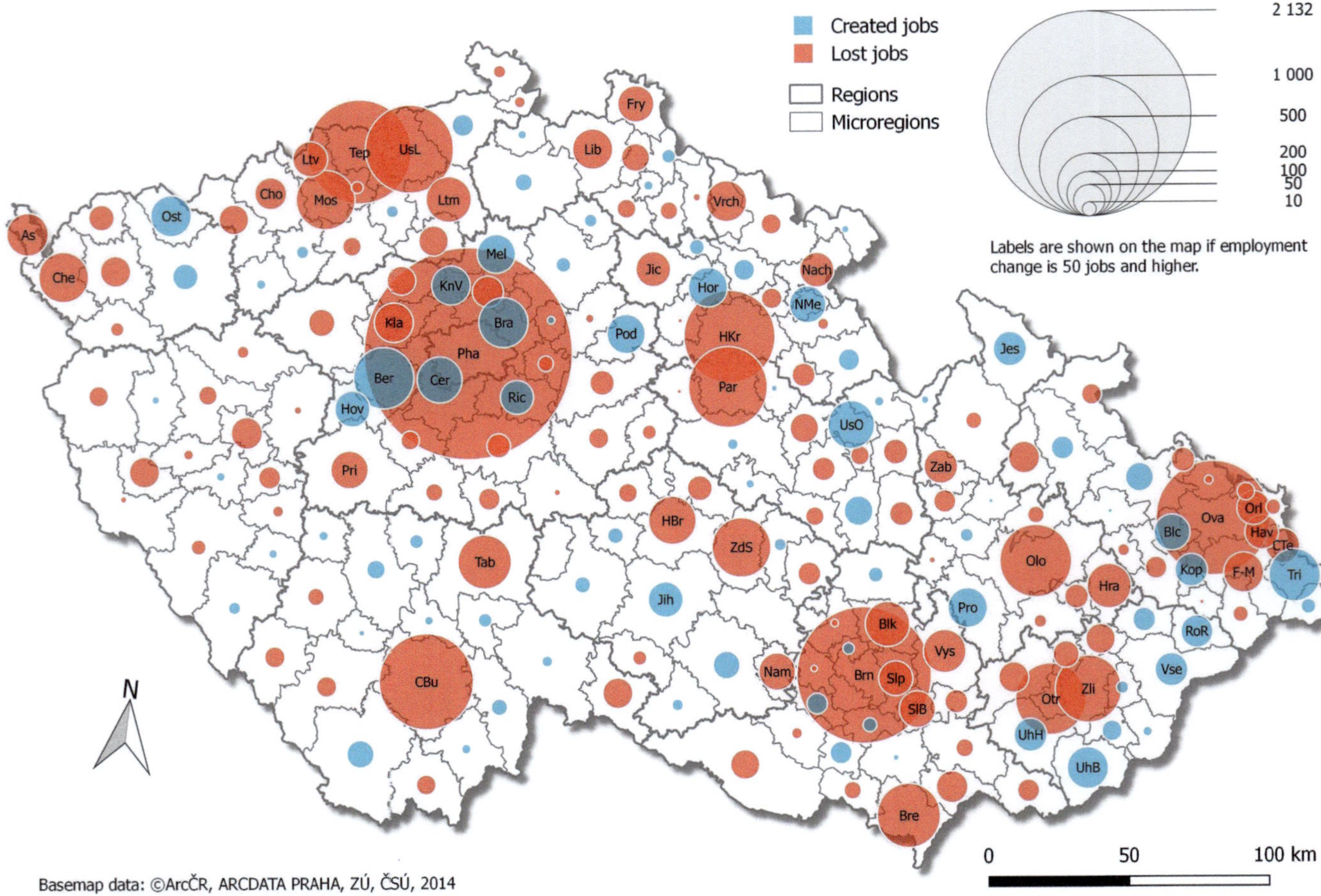

Map 10.3 Employment change in creative industries between 2009 and 2013 at the micro-regional level (Source: CSO 2009; 2013; the authors)

The cores of metropolitan regions (slightly) gained employment exceeding the national average only in the KIBS sector. Somewhat different results were recorded in the hinterlands of metropolitan regions. Except for the other activities sector, all creative industries declined more significantly than in the cores of metropolitan regions, and most visibly in film, TV and video and even in printing. The decline in employment in creative industries in these regions was compensated for by growth in manufacturing employment and high-tech manufacturing employment. To summarise, the development of other metropolitan regions is in line with the findings of Slach et al. (2016), who pointed out that metropolitan regions performed worse in the post-crisis period than most other Czech regions. At the same time, the development in the hinterland of metropolitan regions indicates a tendency towards re-industrialisation rather than tertiarisation.

The Prague metropolitan region, which cannot be overlooked for its hegemonic position, deserves separate attention. Despite general expectations, a core of the metropolitan region has shown a sharp increase in manufacturing industry employment while high-tech manufacturing employment declined. A totally opposite trend was identified in Prague's hinterland, where people lost jobs in manufacturing (mostly high-tech) industries (the biggest decline in Czechia), and even in KIBS, while employment in creative industries increased. Most creative industries grew in the hinterland and declined in the core with no major differences between symbolic and synthetic sectors.

As for the changes in spatial distribution of value added (Table 10.4), the following trends were identified. Most importantly, the spatial concentration of creative industries value added significantly increased. The Prague metropolitan region grew at the cost of all other metropolitan and non-metropolitan regions. In the urban core, employment grew more slowly than value added, while the opposite trend was observed in the hinterland of Prague. A virtually identical trend also occurred (to a lesser intensity) in other metropolitan regions. In the non-metropolitan regions, the decline of value added was lower than that of employment. The cores of metropolitan regions thus created jobs with lower value added, which can be attributed to higher labour costs. The Prague metropolitan region differed most significantly from the rest of Czechia in a dynamic increase of value added in publishing activities, especially in the hinterland of this region.

The change of value added in the motion picture and video sector confirmed its predominantly urban nature, since the decline reached the lowest values. An interesting phenomenon was noticed in the advertising industry. The growth of value added in the core of the Prague metropolitan region was greater than the decline in employment, and, on the contrary, the growth in employment in the hinterland did not correspond with the increase in value added.

10.5 Conclusion

This chapter aimed to evaluate the development of creative industries in the post-crisis period between 2009 and 2013 in Czechia. These industries underwent a number of changes. Firstly, at the sectoral level, their share in both value added and employment were declining. The sectors most affected by the crisis included the motion picture and TV industries. Simultaneously, the job growth in architecture and engineering and other activities was noticed (however, it is necessary to view these results with a certain degree of caution). Particularly, in the case of architecture and engineering, it can be justifiably assumed that the growth was primarily driven by technical expertise and less so by architecture.

Secondly, at the regional level, the rate of spatial concentration of employment and value added increased in favour of Prague, which was, however, primarily due to the insignificant decrease of the observed values in other metropolitan regions. Apart from Prague, the core and hinterland of metropolitan regions, rather than the non-metropolitan regions, surprisingly became the main losers in the post-crisis development of the creative industries. On the other hand, the decrease in these regions was made up for by an increase in high-tech manufacturing.

Thirdly, at the micro-regional level, we register a de-concentration trend of the creative industries in the Prague metropolitan region, but this trend is still relatively weak. Development in other micro-regions took different forms. Apparently, the most obvious trend was the close interconnection between the dynamics of regional economic growth (in particular due to increased specialisation) and the increase or moderate stagnation of the creative industries. It has turned out that the creative industries in Czechia are highly regionalised. Furthermore, micro-regions with a high degree of specialisation in mature industries (printing) reached an above-average rate of employment growth in the creative industries. Finally, in the structurally affected regions, which already had a low proportion of creative industries, the decline in employment and value added reached above-average values.

Fourthly, when compared with the scenarios presented by Pratt (2009), we may conclude that, in the context of Czechia, we have noted a predominantly 'dependent scenario', because the development of creative industries was copying or even 'overtaking' the development of the economy as a whole. The sectoral trajectory of other activities seems to be a partial manifestation of the second scenario, which most likely absorbed some jobs that were lost in other creative industries, especially at the core of the Prague metropolitan region. However, there is a need for closer examination of the nature and type of these new jobs. There is virtually no empirical proof for the third scenario.

All in all, the obtained results suggest that to understand the nature, development, role and importance of the creative industries it is essential to consistently integrate the context in which they operate (see Comunian et al. 2014). With some exaggeration, it can be stated that (market-oriented) creative industries in the context of Czechia as a dependent market economy are largely also market-dependent in terms of the limited ability to generate products and innovations that would create new markets. That in the conditions where demand for such services by the economy controlled by TNCs is relatively weak, limits, for now, their potential contribution to generating the growth and prosperity. Therefore, an analysis of the development of creative industries in the post-crisis period indicates that the expectations outlined in the Introduction are really rather more wishful thinking than a reflection of reality. For the implementation of the first policies to foster the creative industries in Czechia (see Marková 2014), to ignore their specific nature could be a mistake.

Acknowledgments The authors acknowledge support from the Czech Science Foundation through a research grant 'Economic performance of Central-European non-metropolitan regions' (15-03207S).

References

Artadi V, Sala-i-Martin X (2004) The global competitiveness index. In: World economic forum global competitiveness report 2004–2005. Palgrave MacMillan, Hampshire, pp 51–80
Bathelt H, Glückler J (2012) Wirtschaftsgeographie: ökonomische Beziehungen in räumlicher Perspektive. UTB, Stuttgart
Bathelt H, Malmberg A, Maskell P (2004) Clusters and knowledge: local buzz, global pipelines and the process of knowledge creation. Prog Hum Geogr 28(1):31–56
Beaverstock JV, Smith RG, Taylor PJ (1999) A roster of world cities. Cities 16(6):445–458
Blažek J, Csank P (2007) Nová fáze regionálního rozvoje v ČR? Sociologický časopis 43:945–965
Blažek J, Uhlíř D (2002) Teorie regionálního rozvoje: nástin, kritika, klasifikace. Karolinum, Praha
Bohle D, Greskovits B (2007) Neoliberalism, embedded neoliberalism and neocorporatism: towards transnational capitalism in Central-Eastern Europe. West Eur Polit 30(3):443–466
Boix R (2013) Creative industries in Spain: the case of printing and publishing. In: Lazzeretti L (ed.) Creative Industries and Innovation in Europe: Concepts, Measures and Comparative Case Studies. Routledge, London, pp. 65–85
Bundesministerium für Wirtschaft und Technologie (2014) Monitoring zu ausgewählten wirtschaftlichen Eckdaten der Kultur und Kreativwirtschaft 2013. Bundesministerium für Wirtschaft und Technologie, Berlin
Campbell P (2014) Imaginary success?—the contentious ascendance of creativity. Eur Plan Stud 22(5):995–1009
Čermák Z, Hampl M, Müller M (2009) Současné tendence vývoje metropolitních areálů v Česku: dochází k významnému obratu? Geografie 114(1): 37–51
Christopherson S (2002) Project work in context: regulatory change and the new geography of media. Environ Plann A 34(11):2003–2015
Christopherson S (2004) The divergent worlds of new media: how policy shapes work in the creative economy. Rev Policy Res 21(4):543–558
Cikánek M (2009) Kreativní průmysly. Příležitost pro novou ekonomiku. Institut umění - Divadelní ústav, Praha
Cikánek M (2011) Kreativní průmysly: cesta ze země montoven a překladišť. Disk 37:151–160
Comunian R, Chapain C, Clifton N (2014) Creative industries & creative policies: a European perspective? City Cult Soc 2(5):51–53

Cooke P, De Propris L (2011) A policy agenda for EU smart growth: the role of creative and cultural industries. Policy Stud 32(4):365–375
Cruz SS, Teixeira AA (2015) The neglected heterogeneity of spatial agglomeration and co-location patterns of creative employment: evidence from Portugal. Ann Reg Sci 54 (1):143–177
Czech Statistical Office (2009) Register of economic subjects in selected production industries. Internal database of the Czech Statistical Office, Prague
Czech Statistical Office (2010) Register of economic subjects in selected production industries. Internal database of the Czech Statistical Office, Prague
Czech Statistical Office (2013) Register of economic subjects in selected production industries. Internal database of the Czech Statistical Office, Prague
David Q, Peeters D, Van Hamme G, Vandermotten C (2013) Is bigger better? Economic performances of European cities, 1960–2009. Cities 35:237–254
De Propris L (2013) How are creative industries weathering the crisis? Cambridge journal of regions. Econ Soc 6(1):23–35
Drahokoupil J (2007) Analysing the capitalist state in post-socialism: towards the Porterian workfare postnational regime. Int J Urban Reg Res 31(2):401–424
Drahokoupil J, Myant M (2015) Putting comparative capitalisms research in its place: varieties of capitalism in transition economies. In: Ebenau M, Bruff I, May C (eds) New directions in comparative capitalisms research. Palgrave Macmillan UK, Hampshire, pp 155–171
Duranton G, Puga D (2000) Diversity and specialisation in cities: why, where and when does it matter? Urban Stud 37(3):533–555
Duranton G, Puga D (2001) Nursery cities: urban diversity, process innovation, and the life cycle of products. Am Econ Rev 91:1454–1477
EU Commission (2010) Green paper: unlocking the potential of cultural and creative industries. COM (2010), European commission, Brussels
Glaeser EL, Kallal HD, Scheinkman JA, Shleifer A (1991) Growth in cities (No. w3787). National Bureau of Economic Research, Cambridge
Glaeser EL, Henderson V, Inman RP (2000) The future of urban research: nonmarket interactions. In: Brookings-Wharton papers on urban affairs. Brookings Institution Press, Washington, DC, pp 101–149
Gong H, Hassink R (2016) What drives the geographies of creative industries? From literature review to research agenda. Pap Innov Stud 259:1–39
Grabher G (1994) Lob der Verschwendung: Redundanz in der Regionalentwicklung; ein sozioökonomisches Plädoyer. Edition sigma, Berlin
Grabher G (2002) The project ecology of advertising: tasks, talents and teams. Reg Stud 36 (3):245–262
Grabher G (2004) Learning in projects, remembering in networks? Communality, sociality, and connectivity in project ecologies. Eur Urban Reg Stud 11(2):103–123
Grabher G, Stark D (1997) Organizing diversity: evolutionary theory, network analysis and postsocialism. Reg Stud 31(5):533–544
Grodach C, Seman M (2013) The cultural economy in recession: examining the US experience. Cities 33:15–28
Hall PA, Soskice D (eds) (2001) Varieties of capitalism. Oxford University Press, New York
Hampl M (2005) Geografická organizace společnosti v České republice: transformační procesy a jejich obecný kontext. DemoArt, Praha
Heidenreich M (2003) Regional inequalities in the enlarged Europe. J Eur Soc Policy 13 (4):313–333
Henderson V (1997) Medium size cities. Reg Sci Urban Econ 27(6):583–612
Hlaváček P, Koutský J (2011) The polarization tendencies in localization of foreign direct investments in the Czech Republic. Paper presented at 10th International Conference of Liberec Economic Forum, Liberec, 19–20 September 2011

Keeble D, Nachum L (2002) Why do business service firms cluster? Small consultancies, clustering and decentralization in London and southern England. Trans Inst Br Geogr 27 (1):67–90
Kloudová J, Chwaszcz O (2014) The analysis of the creative industry linked in connection with the economic development. E+ M Ekonomie a Manag 12(1):32–42
Kraft S, Halás M, Vančura M (2014) The delimitation of urban hinterlands based on transport flows: a case study of regional capitals in the Czech Republic. Moravian Geogr Rep 22:24–32
Krätke S (2002) Medienstadt. Urbane Cluster und globale Zentren der Kulturproduktion. Leske & Budrich, Opladen
Krätke S (2007) Metropolisation of the European economic territory as a consequence of increasing specialisation of urban agglomerations in the knowledge economy. Eur Plan Stud 15 (1):1–27
Lazzeretti L, Boix R, Capone F (2008) Do creative industries cluster? Mapping creative local production systems in Italy and Spain. Ind Innov 15(5):549–567
Lee N (2014) The creative industries and urban economic growth in the UK. Environ Plann A 46 (2):455–470
Lorenzen M, Frederiksen L (2008) Why do cultural industries cluster? Localization, urbanization, products and projects. In: Cooke P, Lazzaretti R (eds) Creative cities, cultural clusters and local economic development. Edward Elgar, Cheltenham, pp 155–179
Maier K, Franke D (2015) Trendy prostorové sociálně-ekonomické polarizace v Česku 2001–2011. Sociologický časopis 51(1):89–123
Marková B (2014) Creative clusters in the Czech Republic—strategy for local development or fashionable concept? Moravian Geogr Rep 22(1):44–50
Méndez R, Sánchez-Moral S, Malfeito J (2015) Employment changes in knowledge-based industries in large urban areas of Spain: impact of the economic crisis and austerity policies. Environ Plann C Gov Policy 00:1–18
Mossig I (2005) Die Branchen der Kulturökonomie als Untersuchungsgegenstand der Wirtschaftsgeographie. Zeitschrift für Wirtschaftsgeographie 49(2):99–112
Müller K, Rammer C, Trüby J (2009) The role of creative industries in industrial innovation. Innovations 11(2):148–168
Musil J (1993) Changing urban systems in post-communist societies in Central Europe: analysis and prediction. Urban Stud 30(6):899–905
Musil J, Müller J (2008) Vnitřní periferie v České republice jako mechanismus sociální exkluze. Sociologický časopis 44(2):321–348
Myant M, Drahokoupil J (2012) International Integration, Varieties of Capitalism and Resilience to Crisis in Transition Economies. Europe-Asia Studies 64(1): 1–33
Nachum L, Keeble D (2003) Neo-Marshallian clusters and global networks: the linkages of media firms in Central London. Long Range Plan 36(5):459–480
NIPOS (2011) Výsledky účtu kultury ČR za rok 2011. Retrieved from: http://www.czso.cz/csu/2013edicniplan.nsf/p/331813
Nölke A, Vliegenthart A (2009) Enlarging the varieties of capitalism: the emergence of dependent market economies in East Central Europe. World Politics 61(4):670–702
Novák J, Netrdová P (2011) Prostorové vzorce sociálně-ekonomické diferenciace obcí v České republice. Sociologický časopis 47(1):717–744
Novotný J, Blažek J, Květoň V (2016) The anatomy of difference: comprehending the evolutionary dynamics of economic and spatial structure in the Austrian and Czech economies. Eur Plan Stud 24(2):788–808
OECD (2008) Redefining urban: a new way to measure metropolitan regions. Retrieved from http://www.oecd.org/regional/redefiningurbananewwaytomeasuremetropolitanareas.htm
Petrakos G (2001) Patterns of regional inequality in transition economies. Eur Plan Stud 9 (3):359–383
Phelps NA, Ozawa T (2003) Contrasts in agglomeration: proto-industrial, industrial and post-industrial forms compared. Prog Hum Geogr 27(5):583–604

Power D (2003) The Nordic 'cultural industries': a cross-national assessment of the place of the cultural Iindustries in Denmark, Finland, Norway and Sweden. Geografiska Annaler Ser B Hum Geogr 85(3):167–180
Power D, Nielsen T (2010) Priority sector report: creative and cultural industries. European cluster observatory. Center for Strategy and Competitivness, Stockholm
Pratt AC (2005) Cultural industries and public policy: an oxymoron? Int J Cult Policy 11(1):31–44
Pratt AC (2009) The creative and cultural economy and the recession. Geoforum 40(4):495–496
Pratt AC, Hutton TA (2013) Reconceptualising the relationship between the creative economy and the city: learning from the financial crisis. Cities 33:86–95
Rehák Š, Maier G, Buček M (2013) The spatial pattern of creative industries in a transformation economy: the case of Slovakia. Mitteilungen der Österreichischen Geographischen Gesellschaft 55:195–219
Reimer S (2009) Geographies of production II: fashion, creativity and fragmented labour. Prog Hum Geogr 33:65–73
Rozentale I (2014) Creative industries during economic recession: the case of Riga. Reg Stud Reg Sci 1(1):329–335
Rozentale I, Lavanga M (2014) The "universal" characteristics of creative industries revisited: the case of Riga. City Cul Soc 5(2):55–64
Rumpel P, Slach O, Koutský J (2010) Creative industries in spatial perspective in the old industrial Moravian-Silesian region. E+ M Ekonomie Manag 8(4):30–46
Schumpeter JA (1911) Theorie der wirtschaftlichen Entwicklung. Duncker & Humblot, Leipzig. English edition: Schumpeter JA (1934) The theory of economic development: an inquiry into profits, capital, credit, interest, and the business cycle. (trans: Opie R). Transaction publishers, New Brunswick
Scott AJ (2000) The cultural economy of cities. Essays on the geography of image-producing industries. Sage, London
Scott AJ (2006) Creative cities: conceptual issues and policy questions. J Urban Aff 28(1):1–17
Scott AJ (2009) Human capital resources and requirements across the metropolitan hierarchy of the USA. J Econ Geogr 9:207–226
Slach O, Koutský J, Novotný J, Ženka J (2013) Creative industries in the Czech Republic: a spatial perspective. E+ M Ekonomie a Manag 14(4):14–29
Slach O, Ženka J, Koutský J (2016) How resilient are industrial regions in Czechia?. Paper presented at the 6th Halle Forum on Urban Economic Growth, Institute for Economic Research, Halle, 7–8 May 2016
Staber U (2001) The structure of networks in industrial districts. Int J Urban Reg Res 25 (3):537–552
Stam E, De Jong JP, Marlet G (2008) Creative industries in the Netherlands: structure, development, innovativeness and effects on urban growth. Geografiska Annaler Ser B Hum Geogr 90 (2):119–132
Storper M (1997) The regional world: territorial development in a global economy. Guilford Press, New York
TERA (2014) The economic contribution of the creative industries to EU GDP and employment. TERA, Paris
Tomczak P, Stachowiak K (2015) Location patterns and location factors in cultural and creative industries. Quaestiones Geographicae 34(2):7–27
Turok I (2004) Cities, regions and competitiveness. Reg Stud 38(9):1069–1083
Verganti R (2006) Innovating through design. Harv Bus Rev 84(12):1–9
Vernon R (1966) International investment and international trade in the product cycle. Q J Econ 80 (2):190–207
Ženka J, Novotný J, Slach O, Ivan I (2015) Spatial distribution of knowledge-intensive business services in a small post-communist economy. J Knowl Econ 7(4):1000–1020

Chapter 11
Creativity at the European Periphery: Spatial Distribution and Developmental Implications in the Ljubljana Region

Jani Kozina and David Bole

11.1 Introduction

Creativity is gaining importance in the path to establishing a postindustrial knowledge-based society. Contemporary development paradigms often mention creativity as the key factor for economic growth and development trajectories (Hall 2000; McGranahan and Wojan 2007). The main actors in this new post-Fordist production scheme are hypothesised to be creative and knowledge workers, usually referred to as the "creative class", and companies that are active in the so-called creative industries. Increasingly, "creative cities and regions" around the globe have been focusing on these two concepts in pursuing their economic development or regeneration strategies (Chapain et al. 2013). The main logic behind this is to establish a suitable and favourable environment for these kinds of agents and consequently utilise their creative potential in order to boost economic development, competitiveness, social cohesion, urban regeneration, and wellbeing for local residents and newcomers. On the one hand, cities and regions that are more successful in meeting the required criteria of contemporary production-consumption schemes benefit directly and indirectly through a number of spillover effects such as development of new products and services, better management of human resources, improvement of organisational processes, and support branding and communication strategies (Kern 2015). On the other hand, places that are not capable of satisfying the needs and demands of contemporary creative agents suffer from a lack of new ideas and added value, which cause them to lag behind and creates various economic, social, and structural problems.

J. Kozina (✉) • D. Bole
Research Centre of the Slovenian Academy of Sciences and Arts, Anton Melik Geographical Institute, Ljubljana, Slovenia
e-mail: jani.kozina@zrc-sazu.si; david.bole@zrc-sazu.si

© Springer International Publishing AG 2017

C. Chapain, T. Stryjakiewicz (eds.), *Creative Industries in Europe*,
DOI 10.1007/978-3-319-56497-5_11

No matter which concept of creativity we look at, the only general consensus in the vast and growing literature on urban and regional development seems to be its spatial unevenness (see Florida 2002; McGranahan and Wojan 2007; Boschma and Fritsch 2009; Clifton and Cooke 2009; Power and Nielsen 2010; Bakhshi et al. 2015). Everything else—from the definitions, the variables explaining the locational patterns, and the causal mechanisms related to the different socio-economic outcomes, on the one hand (see Musterd and Murie 2010), to the proposed and derived policy instruments, on the other hand (see Musterd and Kovács 2013)—is a matter of extensive scientific and political discussion. First, numerous studies have tried to investigate the creative class and creative industries separately; however there is still a lack of comparative research. It is important to study their interrelational dynamics based on empirical investigation (e.g. Bontje and Musterd 2009; Comunian et al. 2011) in addition to theoretical conceptualisation (Pratt 2008). Paradoxically, the political affirmation of these two concepts has highlighted the need for a common approach (DCMS 2011), although academic circles are still largely ignoring it. Second, the contemporary discourse on creativity, which started in 1994 in Australia, followed by its adoption in the UK and in North America, has now permeated the whole of Europe, Asia, and other parts of the world (Chapain et al. 2013). However, the majority of these current studies still largely concentrate on advanced capitalist economies in Australia, North America, and older European Union member states. New member states and candidate countries in Central and Eastern Europe, the Baltic, and the Balkans remain a predominantly grey area. Third, to understand the geography of the creative economies and to formulate supportive policies, it is necessary to analyse creativity at a sub-regional level (Clifton 2008). An analysis with a higher degree of spatial resolution allows the concentration of a particular economic activity to be more accurately identified. Regions may contain greater variations in economic characteristics within them (e.g. urban vs. rural) than between them, and so comparisons between regions can potentially be misleading (Bakhshi et al. 2015).

The main aim of this chapter is to analyse the spatial distribution of both the creative class and creative industries together in order to identify potential overlap and derive developmental implications using the case study of the Ljubljana region in Slovenia. This is an urban region that is considered peripheral in scholarly and policy literature; it is located outside the main European development areas and corridors such as the "European pentagon" or the area delimitated by London, Hamburg, Munich, Milan, and Paris—that is, the European core with approximately 14% of the EU27 area, 32% of its population, and 43% of its GDP (ESPON 2005). At the same time, it is part of the post-socialist countries in Central Europe, where "regions play a negligible role when world city hierarchies are analysed" (Lang et al. 2015). The main research goal is to show that, just as in "world" or "alpha" cities, creativity also plays an important developmental role in "peripheral", "second-tier", or "beta cities" (Rekers 2012) and is neither unique nor exclusive to larger and hierarchically more central cities such as New York, London, Paris, and Tokyo. In addition, a further step is taken to discover the variability of creativity within the urban region, especially from the urban versus rural or city versus countryside perspective. To this end, we analysed spatial patterns of the creative class and creative industries in the Ljubljana region based on data from the Statistical Register of Employment from the Statistical Office of the Republic of Slovenia and compared this with data from previously investigated regions of western and northern European countries (e.g. Clifton and Cooke 2009; Boschma and Fritsch 2009; DCMS 2011). The developmental implications of creativity in the Ljubljana region were examined by applying statistical correlation analysis between creativity-related factors and indicators of economic development.

11.2 The Creative Class in Europe

The notion of the creative class has been largely popularised in academic and especially policy-making circles since the works of Florida (2002, 2005, 2008), Mellander and Florida (2011), and Mellander et al. (2014). According to this theory, the creative class is composed of people whose job is to create meaningful new forms of creativity. All members of the creative class share a common creative ethos that values creativity, individuality, difference, and merit. For the members of the creative class, every aspect and every manifestation of creativity—technological, cultural, and economic—is interlinked and inseparable. However, Florida makes the further distinction of three subgroups called the creative core, creative professionals, and bohemians. Members of the creative core are those whose economic function is to create new ideas, new technology, and/or new creative content. These individuals are chiefly found in science and engineering, architecture and design, education, arts, music, and entertainment. Creative professionals are those that work in business and finance, law, healthcare, and related fields. They engage in complex problem-solving that involves a great deal of independent judgement and requires high levels of education. By doing this, they are providing support to the creative core to innovate. Bohemians are engaged in cultural and artistic occupations. Their role is twofold. On the one hand, they are a part of the creative class, and, on the other hand, they are a sign of an urban culture of tolerance; thus, they play a key role in attracting the two other categories of the creative class.

The basic creative class concept focuses on three related elements (or the 3Ts): a good "people climate"—with a high level of tolerance, diversity, and amenities (*Tolerance*)—attracts and retains creative and talented people that are highly mobile (*Talent*), who, in turn, fertilise the ground for a competitive business climate (*Technology*), and, finally, a good and competitive "business climate" brings about economic growth. Florida also turned the classical economic theory on its head. Instead of the previously well-established perception that people follow jobs, he proclaimed that jobs follow people. Florida's ideas are supported in the policy-making circles and by some researchers, on the one hand, but have also been seriously criticised on the other hand (for the latest review, see Nathan 2015).

A number of other analyses attempted to reproduce Florida's results in Europe and can be divided into two main strands. The first one primarily originates from the large-scale European research project entitled *Technology, Talent and Tolerance in European Cities: A Comparative Analysis*[1] (e.g. Marlet and van Woerkens 2007; Clifton 2008; Rutten and Gelissen 2008; Boschma and Fritsch 2009; Andersen et al. 2010). Although most found a positive association between some technology businesses, human capital, tolerance, and urban growth, few could reproduce the 3Ts framework (Nathan 2015). The other group of research comes from another large-scale European research project called ACRE, which stands for Accommodating Creative Knowledge[2] (see Musterd and Murie 2010; Musterd and Kovács 2013). These authors adopted a very critical position towards the creative class approach. One of the main conclusions of the project is that the success of cities in the creative economy relies not only on individuals' skills but more broadly on a combination of pathways, place, personal networks, and policies (the 4Ps).

However, despite providing substantive contributions to the discussion on the creative class, its geography, and its spatial distribution, this literature still contains some gaps because it is quite biased towards economically and politically central countries with longer capitalist traditions. One of the more recent books, *Creative Class Goes Global* (Mellander et al. 2014), concentrates the discussion based on selected Nordic and Western European countries (Denmark, Norway, Sweden, Finland, Germany, the Netherlands, England, and Wales), North America, and Australasia (Australia, China, and Japan). The ACRE project extended the agenda to all of Europe by adding some metropolitan areas from new member states, such as Budapest, Poznań, Riga, and Sofia. In addition to these, mention can be made of the work of Działek and Murzyn-Kupisz (2013) and van Heerden and Bontje (2014), who focused on the cases of Kraków and Berlin. All of the previous mappings revealed that the creative class is predominantly urban. The rank-size analysis among 444 city regions in eight European countries showed that cities with a population of less than 70,000 and above 1,200,000 are less attractive to the creative class (Lorenzen and Andersen 2007). However, the creative class is concentrated not only in traditional core cities but also in new centres on the city edges and beyond. More specifically, city centres may mainly attract the "creative core", whereas city edges and beyond may be more attractive for "creative professionals" (Bontje and Kepsu 2013). Some authors also found strong evidence of the positive impact of the creative class on rural growth (McGranahan and Wojan 2007; McGranahan et al. 2010), and so there is a role for policy in harnessing the creativity (often latent) in rural economies as a means of improving their competitiveness (Gülümser et al. 2010; Huggins and Clifton 2011).

One objective of the chapter is to address the two research gaps identified in the geography of the creative class: (1) a lack of comparative analysis between central and peripheral territories at the macro-European level (e.g. Western vs. Eastern

[1]http://www.esrc.ac.uk/my-esrc/grants/RES-000-23-0467/read

[2]http://acre.socsci.uva.nl

Europe or "global/alpha" vs. "regional/beta or gamma" urban regions) and (2) a lack of comparative studies between central and peripheral territories at lower scales (e.g. rural vs. urban or city vs. countryside).

11.3 Creative Industries, Creative Class, and Space

The term *creative industries* originates from local and national governments and their interests in better employment, economic growth, and revitalisation of older industrial urban regions. The Department of Culture, Media, and Sport in the UK (DCMS 2011: 6) has defined creative industries as "those industries which have their origin in individual creativity, skill, and talent and which have a potential for wealth and job creation through the generation and exploitation of economic property". In a post-Fordist economy, cultural forms and meanings are increasingly important if not dominant and a place where human culture is becoming a commodity, a good to be sold on the free market. In other words, a greater share of economic activity is associated with selling and marketing services that have collective attributes of semiotic and aesthetic value (Scott 1997). This turn towards the symbolic, sign, or expressive value in the modern world is considered by some authors (Scott 1997, 2001a; Drake 2003) to be a direct consequence of consumerism, which demands more amusement, ornamentation, self-affirmation, and social display. Catering to these consumers demands are the key creative industries sectors that produce and disseminate artistic and/or semiotic goods and services: advertising, architecture, art and the antiques market, crafts, design, designer fashion, film, interactive leisure software, music, performing arts, publishing, software, and television and radio (Drake 2003: 512).

Empirical studies of creative industries are vast (e.g. Cunningham and Higgs 2008; Bontje and Musterd 2009; O'Connor 2010; Power and Nielsen 2010; Comunian et al. 2011). Whereas research on the creative class focuses on the individuals or the "creatives" in the professional or educational sense, creative industries research most often focuses on the companies and institutions or "creative companies mapping". Cunningham and Higgs (2008) summarise the different ways of using empirical research to demonstrate the scale of creative industries. In its first phase, this encompassed merely counting the number of company employees in individual sectors of creative industries, then later also those occupations that were still creative, but did not fall within the classic creative industries sectors. The newest research is said to focus more on individuals and their occupation: the creative economy encompasses creative professions within the creative industries sectors, creative professions outside the creative industries sectors, and also those professions that support the creatives within the creative industries sectors. On the one hand, the common point to all these creative industries mappings is that it focuses on areas of "concentrations": major cities, especially in Northern and Western Europe, and to an extent in Australia and Asia. On the other hand, creative industries research on the "periphery" (Gibson 2010), on the countryside (Bell and Jayne 2010), or in so-called beta cities (Rekers 2012) is rare, even though cultural production is not unique to large cities, but is possible everywhere. As Gibson and Kong (2005: 549) have noted, it has always been present in major cities to a certain extent, yet in some peripheral areas of the UK, the United States, Australia, New Zealand, and many other countries, it has impacted lifestyles and spatial, economic, and social structures just as profoundly as in major global cities. The literature review of research on creative industries (O'Connor 2010) presents a similar picture as the research on the creative class: there is a lack of comparative studies between central and peripheral territories at macro-spatial levels, a lack of comparative studies between central and peripheral territories at lower scales (city vs. rural countryside), and a lack of longitudinal studies.

The same is true of the comparative research between the creative class and creative industries: there is inherently a degree of overlap at least in terms of creative occupations within creative industries. Some authors reject this comparison due to methodological and contextual inconsistencies and believe that Florida's (2002) definition of the creative class is "both crude and politically repugnant" (Markusen et al. 2008: 27). Pratt (2008) writes that, within the narrow context of enterprise and competition, Florida's notion of the creative class could be a good fit, but at the same time notes serious problems with the notion of creative occupations such as empirical issues and problems of definitions. Bontje and Musterd (2009) provided an important theoretical contribution to drawing the concepts of creative industries and the creative class nearer with their opinion that the research sphere must go beyond "Florida-hyping" and "Florida-bashing" and carry out more comparative research that will confirm or disprove the hypothesis of their connectedness. Here, they cite Scott (2006), who suggests using the concept of "creative field", which "may encompass networks of (potential) companies and employees in creative industries; it also does not exclude theories in which agglomeration economies are central; and due to the multi-scalar character of the concept of the creative field, creative class clustering theories also fit with it". It also allows for the theories of amenities and the importance of places of residence of creative individuals to be integrated in it (Bontje and Musterd 2009: 845).

In the context of creative industries in the Ljubljana region, the second objective of this chapter is to address the importance of creative industries in the "periphery" from a (1) macro-regional (European context) and (2) intra-regional (urban vs. rural or city vs. countryside) perspective. In addition, we consider the overlap of creative industries with the creative class to better understand their relatedness in peripheral regions.

11.4 The Context of Peripherality and Creativity in the Ljubljana Region

The Ljubljana region (NUTS 3 level)[3], with the capital city of Ljubljana, comprises 26 municipalities (LAU 2 level)[4] with approximately 537,000 inhabitants. It is characterised by a monocentric spatial structure with Ljubljana's predominance as an employment urban centre and other suburban and rural municipalities in the countryside. In comparison to other European regions, it exceeds the European average in many parameters, yet it is considered a "peripheral region" from the macro-regional European perspective by many views. The ESPON (2005: 117) study classified the Ljubljana region as the lowest, Category 4 metropolitan growth area (or MEGA), which is described as being "often smaller, less competitive, more peripheral". Slovenia as a country has been listed as "peripheral" based on a global trade index (Chase-Dunn et al. 2000). Hall (2005) mentioned Ljubljana as the regional capital in the sense that it is remote and further from global and sub-global cities at the EU periphery. In this way, Ljubljana is a smaller or medium-sized city in central Europe, distant from the European economic and population core (the European pentagon). Ljubljana is also a representative of a post-socialist city because it shares many traits with other cities in former socialist countries, such as massive deindustrialisation after the 1990s, specific economic and spatial development (Musil 2005), which are also common attributes in other Slovenian cities. This representativeness is limited, however, because every country and city has a specific, path-dependent, and evolutionary post-socialist development (Pavlinek 2003; Stryjakiewicz et al. 2014). However, at the national level, in comparison to other Slovenian regions, it is the most developed one and has a smaller share of the industrial sector, whereas market services and public services are more developed. Its favourable geographical position at the crossroads of the 5th and 10th Trans-European Transport Network with the capital as the main centre strengthens its economic, educational, and cultural dominance in the country. The region has a diversified economy and promotes the service sector and R&D sector development and clustering.

However, EUROSTAT data show that the dispersion in regional per capita GDP in Slovenia is the lowest among the new member states. Between 1996 and 2009, it increased by a smaller magnitude in Slovenia than that recorded in the majority of the new member states (Banerjee and Jesenko 2015). The OECD (2011) has suggested that regional disparities in Slovenia are low because of the country's good infrastructural connectedness and its longstanding regional policies, which supported a scattering of industries across its regions. Slovenia's per capita GDP grew at an annual average rate of 4.2% from 1996 to 2008 but fell markedly

---

[3]NUTS 3 level usually corresponds to population thresholds between 150,000 and 800,000 (EUROSTAT 2011).

[4]LAU 2 level (formerly NUTS 5 level) consists of municipalities or equivalent units in the EU member states (EUROSTAT 2011).

thereafter with the onset of the global financial crisis. Adjusted for differences in purchasing power, Slovenia's per capita GDP was around 84% of the EU-27 average in 2010, much above the levels for other new member states (Banerjee and Jesenko 2015).

Historically, art and culture have always played an important role in Slovenia because they were in a way a substitute for a lack of national, political, and government institutions (Murovec et al. 2014).[5] Unfortunately, no national nor regional strategy for creative industries has yet been developed. The Ministry of Culture has offered some measures, but mostly as individual initiatives or projects (Bradač Hojnik and Rebernik 2014). As in many new member states, there is notable confusion about official policies towards creative or cultural industries because the common perception of cultural policy is that it presents one of the subsidies for the arts and cultural sector from public funds. The concept of market-oriented cultural production (creative industries) is perceived by many cultural producers to be an attempt to undermine the existence of traditional cultural institutions that have been predominantly dependent on state support (Murovec et al. 2014).

In this context, the Regional Development Agency of the Ljubljana Urban Region[6] is seeing new ways and support mechanisms to form a network of creative professionals and enterprises that could eventually evolve into a cluster organisation. In 2011, it established the Regional Creative Economy Centre,[7] the vision of which is to actively contribute to introducing creativity into the public and private sectors by forming a network of creative professionals and enterprises, training courses, workshops, consultations, and other events and by setting up complex value chains. Despite the fact that the Regional Creative Economy Centre aims to promote the creative sector in the entire region and not only in Ljubljana, where it is best developed, recently only three of the 26 municipalities expressed a more serious political commitment to support and develop this sector.

In addition to the lack of provisions at the national level, the local level in Slovenia is weak as well, which opposes the claims of Bontje and Kepsu (2013) that strategies to enhance a creative knowledge economy are well-developed locally or

[5]Slovenia gained its independence in 1991.

[6]In Slovenia there are two levels of government: national and municipal. The regional level is currently defined in various laws that also seek to regulate some activities at the regional level (e.g. in the case of regional policy), but the authorities are still those from the national or local level. The regional development agency is the administrative body that provides administrative activities and technical support to the regional development council (representatives of municipalities, business, and NGOs responsible for preparing the regional development plan, cooperation with other regions, agreement with other parties, territorial dialogue, and monitoring activities at the regional level), the regional council (mayors that ratify the regional development plan and make other decisions at the regional level), and all activities at the regional level. In the case of spatial planning, there is no mandatory body at the regional level; in the case of regional projects, the municipalities involved are expected to make regulations jointly. Altogether there are 12 administrative regions with 12 regional development agencies in Slovenia.

[7]http://www.rcke.si/en/

nationally, contrary to the regional level. In the Ljubljana region, it seems that, despite its lack of any administrative legal power, the Regional Development Agency plays a decisive role in fostering creativity.

11.5 Spatial Patterns of Creativity in the Ljubljana Region

Adding to the current discussions on creativity in central locations, there is a need to explore further issues in relation to the creative class, creative industries, and their intertwining in peripheral areas such as the Ljubljana region. Many authors have emphasised a lack of comparative research between both concepts of creativity and urged the study of their interrelational dynamics based on an empirical investigation (e.g. Bontje and Musterd 2009; Comunian et al. 2011) in addition to the theoretical conceptualisation (Pratt 2008). In order to make this empirical study internationally comparable, we adopted the definition of the creative class from the large-scale European research project Technology, Talent, and Tolerance in European Cities: A Comparative Analysis (see Boschma and Fritsch 2009; Clifton and Cooke 2009) and the definition of creative industries from the UK Department of Culture, Media, and Sport (see DCMS 2011).[8] All comparisons with data from previous studies were based on comparable spatial units in terms of population size. We adapted the data for the Ljubljana region from the Statistical Register of Employment of the Statistical Office of the Republic of Slovenia.[9] To compare both concepts of creativity simultaneously, we analysed employees that were members of the creative class and employees that were active in creative industries in the Ljubljana region in 2011. The numbers revealed that the population of the first group is around 106,000, and approximately 10,400 for the second group. The ratio is almost 10:1. In this way, one can talk about the *wider* and *narrower* definition of creativity. Around 5% of the members of the creative class are active in creative industries, whereas 60% of those active in creative industries can be labelled as the creative class.

[8]The exact definitions of the creative class and creative industries by using the Standard Occupational Classification (SOC) and Standard Industrial Classification (SIC) codes are presented in Appendix 11.1.

[9]Observation units are persons in Slovenia employed through employment contracts and self-employed persons with compulsory social insurance, regardless of whether they work full time or part time and regardless of whether they are on maternity leave, are on leave for care or protection of a child, or are absent from work due to illness or injury or caring for a family member. The number of persons employed excludes persons working under non-employment contracts (contract work), authorship contracts, working students, persons working for direct payment, unpaid family workers, persons employed by employers based in Slovenia that are sent abroad for work or professional training (i.e. employees of Slovenian enterprises at construction sites abroad, etc.), and Slovenian citizens employed by foreign employers abroad (i.e. in neighbouring countries).

11.5.1 The Creative Class: The Wider Definition of Creativity

In 2011, the creative class represented 20% of the inhabitants of the Ljubljana region, which is considerably higher than the national average (15%) and the highest out of all the Slovenian regions. Of this group, about one-third falls into the creative core, about two-thirds into the creative professionals, and a few percent into the bohemian group. An international comparison illustrates that the concentration of the creative workforce in the Ljubljana region is similar to regions of comparable size (NUTS 3 level; 400,000–600,000 inhabitants) in England (see Clifton and Cooke 2009; Fig. 11.1). A similar conclusion can be drawn when comparing the Ljubljana region with regions (NUTS 3 level) in Germany, Norway, Sweden, Finland, and the Netherlands. With the exception of a significant number of Dutch regions and the Greater Stockholm area, the shares of the creative class in the populations almost never exceed 20% (see Boschma and Fritsch 2009: 402).

A higher concentration of the creative class occurs particularly in the central urban section of the Ljubljana region and the nearby suburban municipalities. The concentration progressively declines with the distance from the centre of the region.

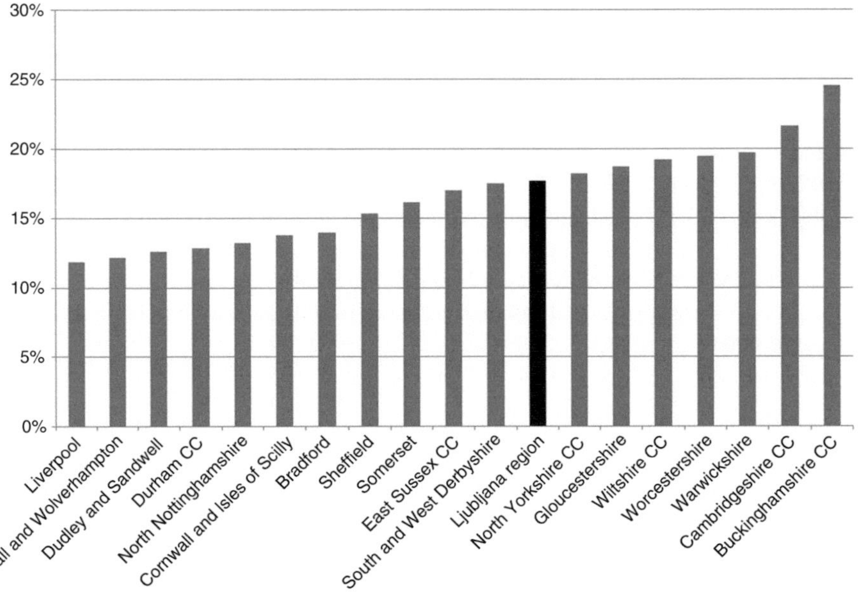

Fig. 11.1 The share of the creative class in the population in the Ljubljana region and comparable English regions (NUTS 3 level; 400,000–600,000 inhabitants) (Source: Own work based on the statistical register of employment; Clifton and Cooke (2009)) (Note: The data on the share of the creative class in the population of the English regions are from 2001, and from 2000 for the Ljubljana region. The data for the English regions were acquired within the large-scale European research project Technology, Talent, and Tolerance in European Cities: A Comparative Analysis (grant no. RES-000-23-0467; see Boschma and Fritsch 2009; Clifton and Cooke 2009))

This concentration is greater with regard to the place of work than to the place of residence (Fig. 11.2a, b). A comparison of how the individual creative subgroups are arranged shows a similar spatial pattern: a great concentration in the heart of the region that decreases towards the rural edges. However, there are certain differences. Members of the creative core and especially bohemians are extremely concentrated in the urban centre compared to the region's rural countryside, whereas creative professionals tend to reside in the suburban areas.

Between 2000 and 2011, the number of people that can be characterised as the creative class grew by almost a third. On the other hand, manufacturing jobs shrunk by almost a third, whereas the number of professions in services and agriculture stayed the same. The global economic crisis of 2008 saw a clear increase in the creative class compared to the other profession groups and it sustained its absolute growth despite increased unemployment. All of the municipalities in the Ljubljana region experienced an increase in the creative workforce during this period; however, this process was least intensive in the region's centre and more intensive on its edges, which coincides with the processes of suburbanisation and periurbanisation (Fig. 11.2c).

11.5.2 Creative Industries: The Narrower Definition of Creativity

In 2011, 1.9% of the inhabitants of the Ljubljana region worked in creative industries, which is somewhat higher than the national average (1.4%). An international comparison shows that the concentration of inhabitants in creative industries in Slovenia is the same as in the UK, at 1.4% (DCMS 2011).[10] A comparison of the individual sectors reveals that the Ljubljana region and the entire country of Slovenia are lagging behind the UK, especially in design, advertising, architecture, software and e-publishing, and digital and entertainment media. Slovenia outperforms the UK only in fashion design, but more extensive investigation revealed that the majority of workers in this sector belong to old textile firms that are in decline. Compared to the UK as whole, the Ljubljana region is better in music, the visual and performing arts, publishing, and television and radio, which is most likely a consequence of the concentration of nationally important cultural institutions, such as Radio-Television of Slovenia, the Slovenian Philharmonic, the Drama theatre, DZS publishers, and others (Fig. 11.3).

The settlement of inhabitants in creative industries in the Ljubljana region shows similar spatial patterns to those of the creative class: a strong concentration in the

[10]The data for both countries take into consideration the number of employees and self-employed in creative industries as a share of total population. The data on employees and self-employed engaged in creative jobs in other industries are omitted from this calculation. The data for Slovenia refer to 2011 and for the UK to 2010.

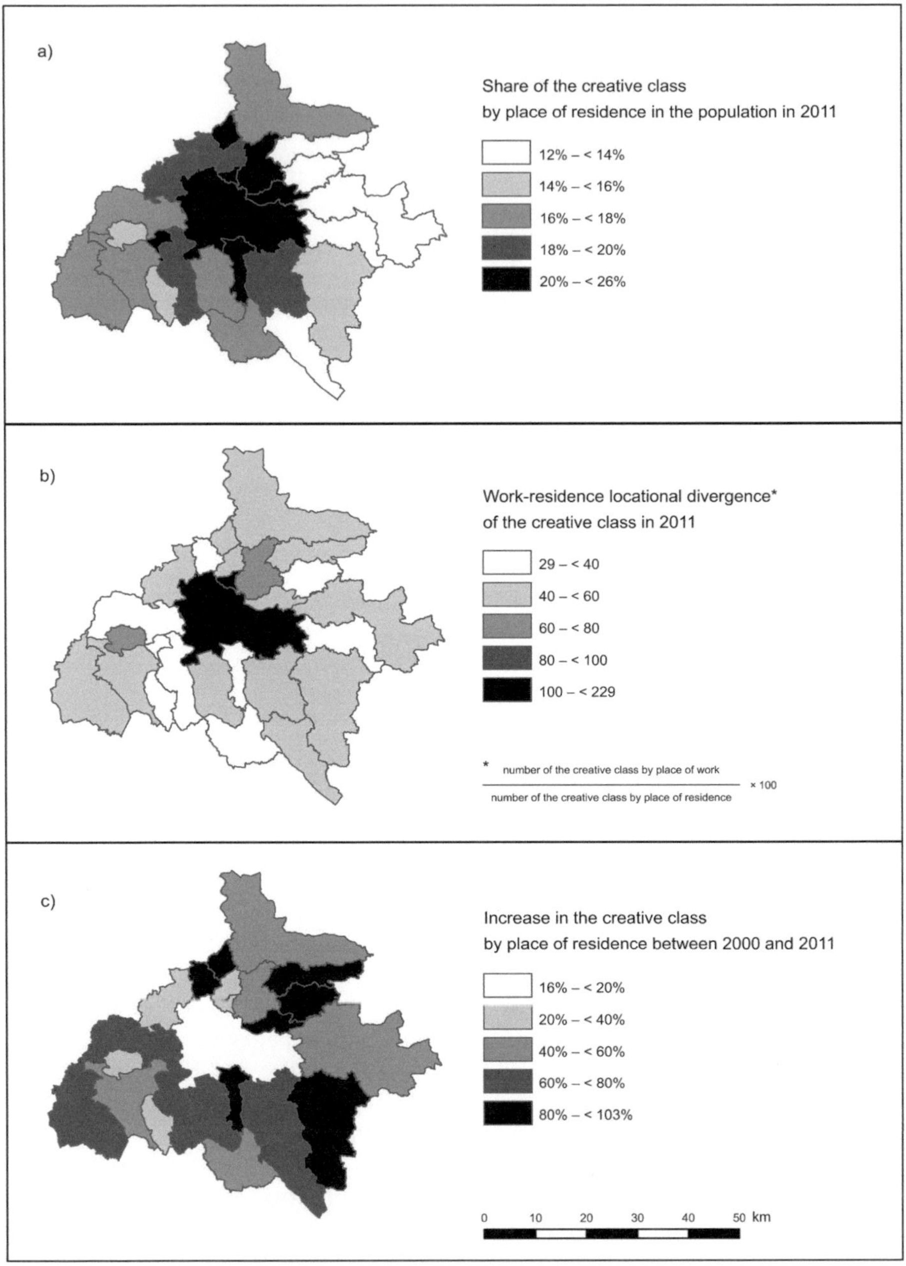

Fig. 11.2 The spatial distribution of the creative class in the Ljubljana region (Source: Own work based on the statistical register of employment)

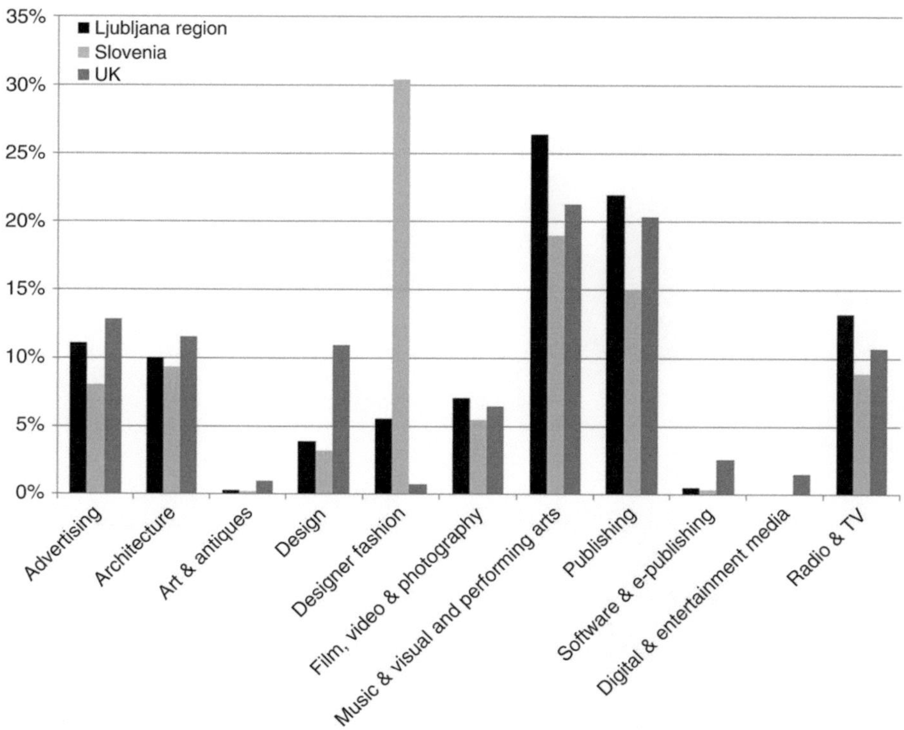

Fig. 11.3 The share of employees in creative industries by place of residence in the Ljubljana region, Slovenia (2011) and UK (2010) (Source: Own work based on the Statistical Register of Employment; DCMS (2011))

central urban area that decreases towards the suburbanised and rural margins. The same is true for the dispersion of jobs, which shows a sharp polarisation and strong daily commuting flows into the heart of the region as opposed to its edges (Fig. 11.4a, b).

However, the temporal implications for the period from 2000 to 2011 show different characteristics in comparison to the creative class. It is evident that the number of inhabitants in creative industries in the Ljubljana region stagnated during that time, and it experienced a steep decline at the scale of the entire country, especially after the economic crisis in 2008. Between 2008 and 2010, the share of inhabitants in creative industries decreased by 13.3% in Slovenia, but only by 0.7% in the Ljubljana region, whereas it grew by 5.0% from 2009 to 2010 in the UK (DCMS 2011). The changes within the region show that the number of inhabitants in creative industries increased only in some suburban and rural municipalities south and north of Ljubljana. The share of this sector decreased everywhere else, including in the urban centre of the region. These trends are not aligned with common suburbanisation and periurbanisation spatial processes. Consequently, they deserve more thorough investigation in future research.

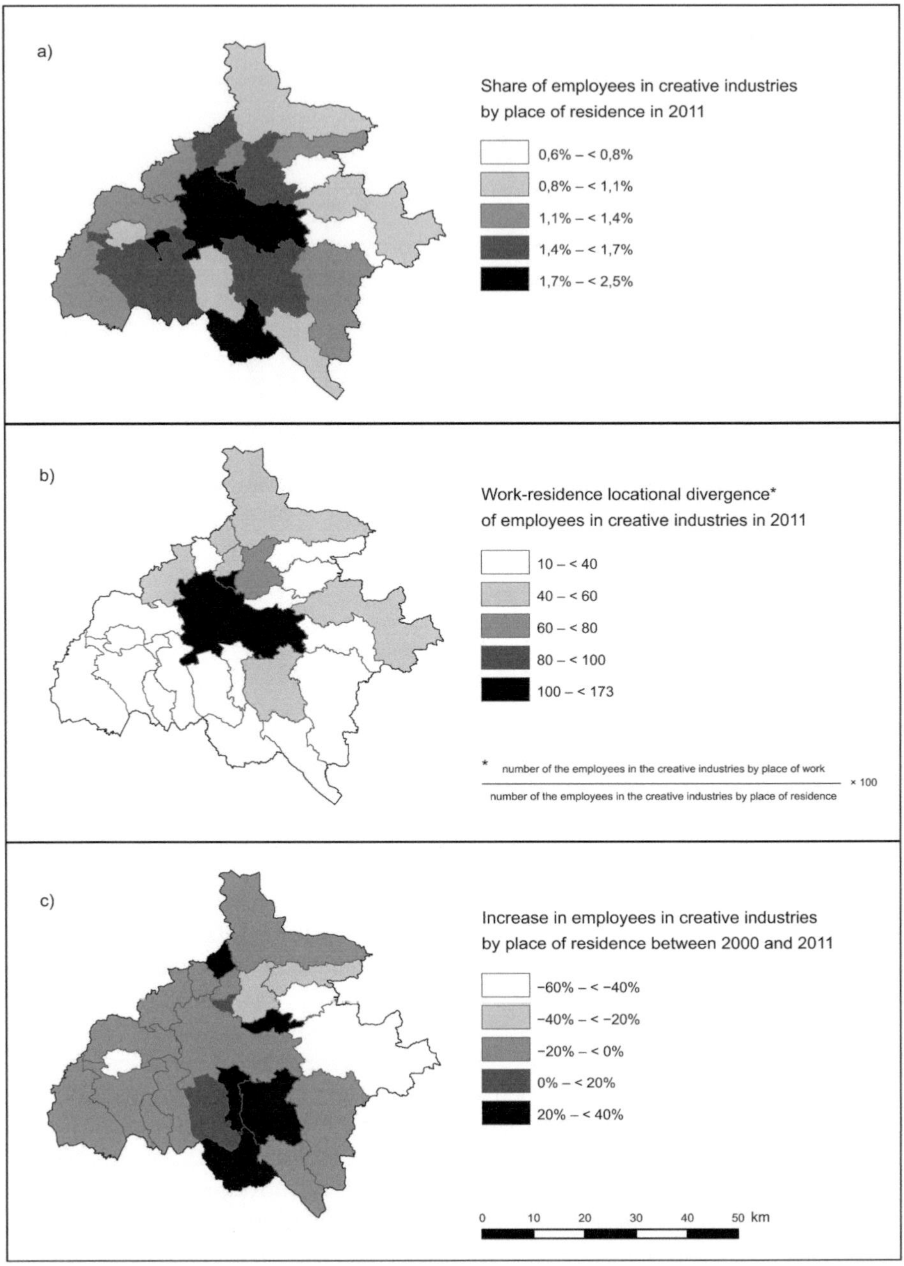

Fig. 11.4 The spatial distribution of employees in creative industries in the Ljubljana region (Source: Own work based on the statistical register of employment)

11.6 Creativity as a Driver of Economic Development in the Ljubljana Region

The comparison of spatial patterns of main creative agents between centrally located regions in Western and Northern European regions, on the one hand, and peripheral regions such as the Ljubljana region, on the other hand, imply that "centrality", urban hierarchy, and city size are not factors for predicting the creative class and creative industries in numbers. In addition, it is important to investigate whether creativity also plays a role in regional economic development in the context of peripherality.

The correlation between creativity and economic development in municipalities of the Ljubljana region was examined using bivariate correlation analysis (*r*). For this purpose, indicators and factors of economic development were formed. The factors were further divided into creativity-related factors and other factors that were added for greater insight into the importance of creativity for economic development in the Ljubljana region (Table 11.1).

The results show that there is an important correlation between the level of creativity in the municipalities in the Ljubljana region and the level of economic development. The contribution of the creative class is significant by the place of residence and place of work. The areas of residence of the creative class have a higher gross taxable income, whereas the areas where the creative class works have a higher added value per employee, higher concentration of workplaces, and higher wages. Creative industries display similar tendencies as the creative class with regard to the place of residence: a strong correlation with gross taxable income and no significant influence on the unemployment rate. There is no significant correlation between the place of work and economic development.

When comparing the correlation between economic development and the factors that are not directly associated with creativity, the education level is the variable that shows the strongest level of correlation with the creative class. In addition, other indicators also show a more or less positive effect on economic development. The only exception to this are other, "non-creative" occupational groups. Of course, this does not mean that they are not important to the economy. People need agricultural products, simple industrial products, and services not only because they also need good accessibility, investment, and a tolerant environment, but more as support for the contemporary post-Fordist economy, whose main driver seems to be creativity according to the correlation analysis. One must also realise that creativity is often an integral component of the production processes in the agricultural, manufacturing, and service sectors. The problem is that the system of measurement implemented cannot make this aspect of creativity visible; this remains a challenge for future methodological investigations. Some correlation coefficients have negative values in regional centres accessibility and road accessibility; the reason for this is in the structure of both of these indicators. Specifically, accessibility increases with a shorter travel time to the destination.

Table 11.1 Coefficients of correlation (r) between creativity and economic development in the Ljubljana region in 2011

Factors		Indicators of economic development				
		Gross taxable income	Unemployment rate	Added value per employee	Workplaces	Wages
Creativity-related factors of economic development	Creative class (A + B + C) _residence	.948**	.075			
	Creative core (A)_residence	.848**	.164			
	Creative professionals (B) _residence	.946**	.012			
	Bohemians (C)_residence	.815**	.120			
	Creative class (A + B + C) _workplace			.474*	.521**	.691**
	Creative core (A)_ workplace			.177	−.128	.327
	Creative professionals (B) _workplace			.510**	.723**	.685**
	Bohemians (C)_workplace			.010	.181	.244
	Employees in creative industries_residence	.803**	.155			
	Employees in creative industries_workplace			−.093	.240	.158

Other factors of economic development	Agricultural occupations_residence	−.779**	−.312			
	Manufacturing occupations_residence	−.754**	.150			
	Service occupations_residence	−.859**	−.011			
	Population	.190	.554 **	.247	.353	.405*
	Population density	.786**	.217	.332	.672**	.450*
	Population growth 2000–2011	.430*	−.286	089	.068	.190
	Education level	.931**	.085	.055	.624**	.258
	Investment	.517**	.321	.531	.801**	.611**
	Internet accessibility	.661**	−.024	.286	.763**	.437*
	Regional centres accessibility	−.593**	−.271	−.335	−.552**	−.498**
	Road accessibility	−.486*	.170	−.291	-.234	−.331
	Public transport accessibility	.677**	−.025	.277	.628**	.408*
	Openness	.498**	.199	−.440*	.101	−.325

The exact definitions and descriptive statistics of the indicators and factors of economic development are presented in Appendix 11.2.
$*p < .05$ (2-tailed), $**p < .01$ (2-tailed)

Finally, it is also important to determine whether there is a correlation between the distribution of the creative class and the creative industries. The results show that there is a very high correlation between both sets of variables. Even though these are two quite different aspects of creativity—with regard to their contents and scope—they are very intertwined with regard to their spatial distribution. The results of the statistical analysis clearly show the association between creativity in the municipalities in the Ljubljana region and their level of economic development. However, additional research is needed to reveal causal mechanisms behind this and to identify factors that hinder the full release of creative potential.

11.7 Discussion and Conclusion: Developmental Potentials for Creativity in the Ljubljana Region

This chapter sought to explain the spatial dynamics of creative economies in the periphery—not only at the European and national levels, but also in the sub-regional context, where there is a lack of comparative research. First, the share of the creative class and employment in the creative industries in Ljubljana paints a picture similar to other much larger and central cities in Europe, implying that "centrality", urban hierarchy, and city size are not factors for predicting the creative class and creative industries in numbers. Second, the results show that creativity is not unique to advanced capitalist economies and that it has a significant developmental role in the Ljubljana region because it can be associated with local economic development. Third, the distribution of creativity at a sub-regional level identified central locations (the city centre and densely populated urban areas) with the greatest concentration of creativity, although there are indications that creativity is turning increasingly towards the suburban and rural hinterland. Fourth, comparing the creative class and creative industries overlap revealed the need to talk about the wider and narrower definition of creativity. Although the creative class is vast and represents a majority within creative industries, the employees in creative industries are fewer and represent only a small share of the creative class.

Although the Ljubljana region is not a "global alpha city region," which is defined as a larger and hierarchically more central region within advanced economies (Rekers 2012), it has a high concentration of the creative class and creative industries. In this aspect, we argue, it is completely comparable to similarly sized regions in most developed countries of Northern and Western Europe. Similar findings were revealed in some other parts of Central and Eastern Europe—for instance in Bratislava, where creative and cultural industries command an unusually high share of the regional employment (Power and Nielsen 2010). The case of the Ljubljana region shows that research on creativity is somewhat biased because it is often based on agglomerative logic, which tends to diminish the role of smaller and peripheral urbanities. Scott (2001b) thinks that the bulk of creative production is taking place in large, metropolitan cities due to agglomerative tendencies that originate from the sole nature of creative production: interdependency amid producers and consumers and incentives among them to locate in close proximity to each other. According to Sassen (1999), the new urban economy is forcing entrepreneurial firms to concentrate in major global cities, which are the only locations where they are able to keep up with the fast pace of trends and ideas. As a result, many smaller and less central urban areas are often ignored in academic research or are merely seen as places from which creative youth migrate to other larger urban areas (Hall and Donald 2013).

The fact that small and peripheral places such as the Ljubljana region have an important creative sector suggests that one needs to look beyond Scott's "agglomerative" logic (2001b) and investigate the factors attracting and retaining the creative class and creative industries in the periphery. We agree with Hall and Donald (2013: 89), who suggest that perhaps the periphery offers a "rich quality of life and an alternative to the fast pace of large cities", with housing affordability and a better natural environment as obvious advantages. Knox and Mayer (2013) emphasise the role of natural amenities, local traditions, and heritage as the main advantages of smaller (or "slower") cities over those located in the European central pentagon area. Perhaps the agglomeration of the creative sector in Ljubljana could partly be explained with its status of a state capital and thus a national cultural and political centre. Such reasoning implies that one needs to look beyond simple attracting and retaining factors of regions and involve their context, pathways, and specificities (Bontje et al. 2011). This suggestion also holds important policy implications for the Ljubljana region and other peripheral areas. Instead of focusing on policies better suited to central and larger regions (such as Florida's 3T concept), peripheral regions could instead focus on maintaining and improving those qualities that give them competitive advantages, such as housing affordability, a natural environment, or good education. This view could be more strongly related to the 4P concept, which highlights the importance of pathways, place, personal networks, and policies in maintaining the creative sector in cities and regions (Musterd and Murie 2010; Musterd and Kovács 2013).

At the sub-metropolitan or intra-regional level, the spatial distribution of the creative class in the Ljubljana region is indicating a slight shift from the urban core towards the suburban and rural hinterland. This is in line with the findings of Bontje and Kepsu (2013) that "creative knowledge workers" and "creative knowledge companies" in some European metropolitan areas such as Amsterdam, Birmingham, Helsinki and Toulouse are concentrated not only in traditional core cities, but also in new centres at the city edges and beyond. However, some differences were noticed when comparing the creative class that signifies stronger suburban and periurban tensions with creative industries that tend to follow this pattern to a limited extent. Dispersion of the creative class towards suburban and rural settlements is typical for all the regions in

Slovenia and probably corresponds to the general settlement tendencies of wanting to live in a calm, green, and sparsely populated environment with good access to larger centres and infrastructure, which has been characteristic of the Ljubljana region and Slovenia for the past few decades (Ravbar 1997; Hočevar et al. 2005). If one looks closely, very similar factors are mentioned at the macro-geographical scale, such as lower costs of living, better natural amenities, and quality of the living environment. In relation to this, McGranahan et al. (2010) provided some empirical evidence by linking outdoor amenities with the creative class concentration in the United States. Nevertheless, our findings do not necessarily mean that central urban locations are less popular as places of residence and work; however, they do imply that "urban creative" policies should be re-examined in the more rural context.

The creativity potential in the Ljubljana region is high, but its release is not being carried out in an adequately planned and systematic manner, even though it already correlates with many indicators of economic development, as is evident from the correlation analysis. One of the first tasks the region should carry out without waiting for the national level is to develop a strategy of creative economy. The newly formed Regional Creative Economy Centre has produced some content that has been included in the Regional Development Program for the Ljubljana Urban Region 2014–2020, but a more comprehensive document is needed with a clearly expressed and precisely defined vision, goals, measures, responsible stakeholders, financing, and expected results. One of the first tasks would be to define which creative occupations/industries have good potential with regard to the critical mass of consumers, tradition, education, and availability of creatives, as well as to decide where these activities should be encouraged (i.e. their locations). The strategy should also focus on the amenities typical for rural and suburban locations that attract creatives (i.e. a natural environment, affordable housing, and good accessibility). Based on previous research, architecture and design were already recognised by state and local authorities as the two sub-branches of further interest for analysis and networking support in Slovenia (Murovec et al. 2014). However, more research is needed to investigate which sectors of the creative economy could offer the most added value, and where within the Ljubljana region; not just in relation to economic aspects, but also in relation to social inclusion, better quality of life, and the growth of other "non-creative" sectors.

Acknowledgement The work for this chapter was partially supported by the Regional Development Agency of the Ljubljana Urban Region, Regional Creative Economy Centre under contract 1479/13-AC entitled Distribution of Creativity in Selected Municipalities of the Ljubljana Urban Region.

Appendices

Appendix 11.1 Definition of the creative class and creative industries using Standard Occupational Classification (SOC) and Standard Industrial Classification (SIC) codes

Creative class			Creative industries		
Group	Standard Occupational Classification (SOC)		Sector	Standard Industrial Classification (SIC)	
	Code	Description		Code	Description
Creative core	211	Physicists, chemists, and related professionals	1. Advertising	73.110	Advertising agencies
	212	Mathematicians, statisticians, and related professionals		73.120	Media representation
	213	Computing professionals	2. Architecture	71.111	Architectural planning
	214	Architects, engineers, and related professionals		71.112	Landscape architecture, urban, and other planning
	221	Life science professionals	3. Arts & antiques	47.782	Retail sale services of commercial art galleries
	222	Health professionals (except nursing)	4. Crafts	No SIC codes to identify these specifically. The majority of businesses are too small to be picked up in business surveys	
	231	College, university, and higher education teaching professionals	5. Design	74.100	Specialised design activities
	232	Secondary education teaching professionals	6. Designer fashion	14.110	Manufacture of leather clothes
	233	Pre-primary, first-stage, and second-stage basic education teaching professionals		14.120	Manufacture of workwear
	234	Special education teaching professionals		14.130	Manufacture of other outerwear
	235	Other teaching professionals		14.140	Manufacture of underwear
	243	Archivists, librarians, and related information professionals		14.190	Manufacture of other wearing apparel and accessories
	244	Social science and related professionals		14.200	Manufacture of articles of fur

(continued)

Appendix 11.1 (continued)

Creative class			Creative industries		
	247	Other public service administrative professionals not elsewhere classified		14.310	Manufacture of knitted and crocheted hosiery
	344	Customs, tax, and related government-associated professionals		14.390	Manufacture of other knitted and crocheted apparel
Creative professionals	1	Legislators, senior officials, and managers		15.120	Manufacture of luggage, handbags and the like, saddlery and harness
	223	Healthcare professionals		15.200	Manufacture of footwear
	241	Business professionals	7. Video, film, and photography	59.110	Motion picture, video, and television programme production activities
	242	Legal professionals		59.120	Motion picture, video, and television programme post-production activities
	31	Physical and engineering science-associated professionals		59.130	Motion picture, video, and television programme distribution activities
	32	Life science and health-associated professionals		59.140	Motion picture projection activities
	341	Finance and sales-associated professionals		74.200	Photographic activities
	342	Business services agents and trade brokers	9 and 10. Music, visual and performing arts	59.200	Sound recording and music publishing activities
	343	Administrative-associated professionals		90.010	Performing arts
	345	Police investigation and related associated professionals		90.020	Support activities to performing arts
	346	Social work-associated professionals		90.030	Artistic creation
Bohemians	245	Writers and creative or performing artists		90.040	Operation of arts facilities
	347	Artistic, entertainment, and sports-associated professionals	11. Publishing	18.110	Printing of newspapers

(continued)

Appendix 11.1 (continued)

Creative class			Creative industries		
	521	Fashion and other models		18.130	Pre-press and pre-media services
				58.110	Book publishing
				58.130	Publishing of newspapers
				58.140	Publishing of journals and periodicals
				58.190	Other publishing activities
				63.910	News agency activities
			8 and 12. Software and e-publishing	18.200	Reproduction of recorded media
				58.290	Other software publishing
			8 and 12. Digital and entertainment media	58.210	Publishing of computer games
			13. Radio and TV	60.100	Radio broadcasting
				60.200	Television programming and broadcasting activities

Appendix 11.2 Descriptive statistics of indicators and factors of economic development in municipalities of the Ljubljana region

	Short name	Description	Mean	Standard deviation
Indicators of economic development	Gross taxable income	Gross taxable income per capita in € in 2010	8937.05	1029.54
	Unemployment rate	Share of registered unemployed persons in the active population by place of residence in 2011	8.28	1.75
	Added value per employee	Added value per employee in € in 2010	32,443.24	6412.76
	Workplaces	Number of jobs per 1000 inhabitants in 2011	288.81	257.39
	Wages	Average annual gross earnings in € in 2011	1395.45	155.87
Creativity-related factors of economic development	Creative class (A + B + C)_residence	Share of the creative class in the active population by place of residence in 2011	41.91	8.14
	Creative core (A)_residence	Share of the creative core in the active population by place of residence in 2011	12.56	2.63
	Creative professionals (B)_residence	Share of creative professionals in the active population by place of residence in 2011	28.10	5.32
	Bohemians (C) _residence	Share of bohemians in the active population by place of residence in 2011	1.24	0.58
	Creative class (A + B + C) _workplace	Share of the creative class in the active population by place of work in 2011	33.62	6.85
	Creative core (A)_ workplace	Share of the creative core in the active population by place of work in 2011	10.03	2.91
	Creative professionals (B)_workplace	Share of creative professionals in the active population by place of work in 2011	22.60	5.33
	Bohemians (C) _workplace	Share of bohemians in the active population by place of work in 2011	0.98	0.51
	Employees in creative industries_residence	Share of employees in creative industries in the active population by place of residence in 2011	3.28	1.03
	Employees in creative industries_workplace	Share of employees in creative industries in the active population by place of work in 2011	2.17	1.03

(continued)

Appendix 11.2 (continued)

	Short name	Description	Mean	Standard deviation
Other factors of economic development	Agricultural occupations_residence	Share of agricultural occupations in the active population by place of residence in 2011	4.38	2.67
	Manufacturing occupations_residence	Share of manufacturing occupations in the active population by place of residence in 2011	10.35	2.45
	Service occupations_residence	Share of service occupations in the active population by place of residence in 2011	41.86	4.30
	Population	Number of inhabitants in 2011	20,508.19	53,539.07
	Population density	Number of inhabitants per km^2 in 2011	184.62	205.45
	Population growth 2000–2011	Index of inhabitants in 2000–2011	117.60	8.81
	Education level	Share of inhabitants with tertiary education in 2011	26.33	5.44
	Investment	Gross fixed capital formation in 2006–2010 (€ in 1000) per 1000 inhabitants	6230.18	6483.66
	Internet accessibility	Number of broadband connections per 1000 households in 2010	583.12	111.64
	Regional centres accessibility	Average travel time by car to the nearest regional centre in 2005	22.65	6.94
	Road accessibility	Average travel time by car to the nearest freeway access in 2005	9.31	5.59
	Public transport accessibility	Share of residents living within a radius of 500 m from public transport stops in 2006	53.00	17.69
	Openness	Share of employees with foreign citizenship in the labour force by place of work in 2011	8.15	5.86

References

Andersen KV, Hansen HK, Isaksen A et al (2010) Nordic city regions in the creative class debate: putting the creative class thesis to a test. Ind Innov 17:215–240. doi:10.1080/13662711003633496

Bakhshi H, Davies J, Freeman A et al (2015) The geography of the UK's creative and high-tech economies. NESTA. https://www.nesta.org.uk/sites/default/files/geography_uks_creative_high-tech_economieswv20151.pdf. Accessed 15 Feb 2016

Banerjee B, Jesenko M (2015) Economic growth and regional disparities in Slovenia. Reg Stud 49:1722–1745. doi:10.1080/00343404.2013.879981

Bell D, Jayne M (2010) The creative countryside: policy and practice in the UK rural cultural economy. J Rural Stud 26:209–218. doi:10.1016/j.jrurstud.2010.01.001

Bontje M, Kepsu K (2013) Creative knowledge strategies for polycentric city-regions. In: Musterd S, Kovács Z (eds) Place-making and policies for competitive cities. Wiley-Blackwell, Oxford, pp 191–208

Bontje M, Musterd S (2009) Creative industries, creative class and competitiveness: expert opinions critically appraised. Geoforum 40:843–852. doi:10.1016/j.geoforum.2009.07.001

Bontje M, Musterd S, Kovács Z et al (2011) Pathways toward European creative-knowledge city-regions. Urban Geogr 32:80–104. doi:10.2747/0272-3638.32.1.80

Boschma RA, Fritsch M (2009) Creative class and regional growth: empirical evidence from seven European countries. Econ Geogr 85:391–423. doi:10.1111/j.1944-8287.2009.01048.x

Bradač Hojnik B, Rebernik M (2014) Insight into creative and cultural industries in Slovenia. Bus Manag Dyn 3:10–15

Chapain C, Clifton N, Comunian R (2013) Understanding creative regions: bridging the gap between global discourses and regional and national contexts. Reg Stud 47:131–134. doi:10.1080/00343404.2013.746441

Chase-Dunn C, Kawano Y, Brewer BD (2000) Trade globalization since 1795: waves of integration in the world-system. Am Sociol Rev 65:77–95

Clifton N (2008) The 'creative class' in the UK: an initial analysis. Geografiska Annaler Ser B Hum Geogr 90:63–82. doi:10.1111/j.1468-0467.2008.00276.x

Clifton N, Cooke P (2009) Creative knowledge workers and location in Europe and North America: a comparative review. Creat Ind J 2:73–89. doi:10.1386/cij.2.1.73/1

Comunian R, Faggian A, Jewell S (2011) Winning and losing in the creative industries: an analysis of creative graduates' career opportunities across creative disciplines. Cult Trends 20:291–308. doi:10.1080/09548963.2011.589710

Cunningham SD, Higgs PL (2008) Creative industries mapping: where have we come from and where are we going? Creat Ind J 1:7–30. doi:10.1386/cij.1.1.7/1

Department for Culture Media and Sport [DCMS] (2011) Creative industries economic estimates. Full statistical release. https://www.gov.uk/government/uploads/system/uploads/attachment_data/file/77959/Creative-Industries-Economic-Estimates-Report-2011-update.pdf. Accessed 11 Dec 2013

Drake G (2003) This place gives me space: place and creativity in the creative industries. Geoforum 34:511–524. doi:10.1016/S0016-7185(03)00029-0

Działek J, Murzyn-Kupisz M (2013) Is Krakow attractive to young members of the creative class? Urban policy implications. In: Wiktor-Mach D, Radwański P (eds) Conference proceedings, the idea of creative city/The urban policy debate, Cracow, 17–18 October 2013. http://eujournal.org/files/journals/1/books/Cracow2013.pdf. Accessed 11 Dec 2015

ESPON (2005) Espon 1.1.1: potentials for polycentric development in Europe. Project report. Nordergio, Stockholm

EUROSTAT (2011) Regions in the European Union. Nomenclature of territorial units for statistics NUTS 2010/EU-27. http://ec.europa.eu/eurostat/documents/3859598/5916917/KS-RA-11-011-EN.PDF. Accessed 21 Feb 2016

Florida R (2002) The rise of the creative class: and how it's transforming work, leisure, community and everyday life. Basic Books, New York

Florida R (2005) The flight of the creative class: the new global competition for talent. HarperCollins, New York

Florida R (2008) Who's your city? How the creative economy is making where you live the most important decision of your life. Basic Books, New York

Gibson C (2010) Guest editorial–creative geographies: tales from the 'margins'. Aust Geogr 41:1–10. doi:10.1080/00049180903535527

Gibson C, Kong L (2005) Cultural economy: a critical review. Prog Hum Geogr 29:514–561. doi:10.1191/0309132505ph567oa

Gülümser AA, Baycan-Levent T, Nijkamp P (2010) Measuring regional creative capacity: a literature review for rural-specific approaches. Eur Plan Stud 18:545–563. doi:10.1080/09654311003593614

Hall P (2000) Creative cities and economic development. Urban Stud 37:639–649. doi:10.1080/00420980050003946

Hall P (2005) The world's urban systems: a European perspective. Glob Urban Dev 1:1–12

Hall HM, Donald B (2013) Clarifying creativity and culture in a small city on the Canadian periphery: challenges and opportunities in Greater Sudbury. In: Lorentzen A, Van Heur B (eds) Cultural political economy of small cities. Routledge, Abingdon, pp 80–94

Hočevar M, Uršič M, Kos D et al (2005) Changing of the Slovene urban system: specific socio-spatial trends and antiurban public values/attitudes. In: Eckardt P (ed) Paths of urban transformations, The European city in transition, vol 5. Peter Lang, Frankfurt am Main, pp 281–300

Huggins R, Clifton N (2011) Competitiveness, creativity, and place-based development. Environ Plann A 43:1341–1362. doi:10.1068/a43559

Kern P (ed) (2015) The smart guide to creative spill-overs. http://www.keanet.eu/wp-content/uploads/SMARTGUIDE-FINAL-PDF.pdf?4f4eb7. Accessed 22 Dec 2015

Knox PL, Mayer H (2013) Europe's internal periphery: small towns in the context of reflexive polycentricity. In: Lorentzen A, Van Heur B (eds) Cultural political economy of small cities. Routledge, Abingdon, pp 142–158

Lang T, Henn S, Sgibnev W et al (2015) Understanding new geographies of Central and Eastern Europe. In: Lang T, Henn S, Sgibnev W et al (eds) Understanding geographies of polarization and peripheralization. Palgrave MacMillan, Basingstoke, pp 1–24

Lorenzen M, Andersen KV (2007) The geography of the European creative class: a rank-size analysis. Working Paper 7. Danish Research Unit for Industrial Dynamics, Frederiksberg

Markusen A, Wassall G, DeNatale D et al (2008) Defining the creative economy: industry and occupational approaches. Econ Dev Q 22:24–45. doi:10.1177/0891242407311862

Marlet G, van Woerkens C (2007) The Dutch creative class and how it fosters urban employment growth. Urban Stud 44:2605–2626. doi:10.1080/00420980701558434

McGranahan DA, Wojan T (2007) Recasting the creative class to examine growth processes in rural and urban counties. Reg Stud 41:197–216. doi:10.1080/00343400600928285

McGranahan DA, Wojan TR, Lambert DM (2010) The rural growth trifecta: outdoor amenities, creative class and entrepreneurial context. J Econ Geogr 11:529–557. doi:10.1093/jeg/lbq007

Mellander C, Florida R (2011) Creativity, talent, and regional wages in Sweden. Ann Reg Sci 46:637–660. doi:10.1007/s00168-009-0354-z

Mellander C, Florida R, Asheim BT et al (eds) (2014) The creative class goes global. Routledge, Abingdon

Murovec N, Kavaš D, Cerar A (2014) Creative industries in Slovenia: clustering, analysis and challenges. In: Petrin T, Břusková P, Bialic-Davendra M (eds) Cluster development in the Czech Republic and Slovenia. Faculty of Economics, Ljubljana, pp 121–146

Musil J (2005) City development in Central and Eastern Europe before 1990: historical context and socialist legacies. In: Hamilton I, Dimitrovska Andrews K, Pichler-Milanovič N (eds) Transformation of cities in Central and Eastern Europe: towards globalisation. United Nations University Press, Tokyo, pp 22–43

Musterd S, Kovács Z (eds) (2013) Place-making and policies for competitive cities. Wiley-Blackwell, Oxford
Musterd S, Murie A (eds) (2010) Making competitive cities. Wiley-Blackwell, Oxford
Nathan M (2015) After Florida: towards an economics of diversity. Eur Urban Reg Stud 22:3–19. doi:10.1177/0969776412463371
O'Connor J (2010) The cultural and creative industries: a literature review. Creativity, Culture and Education, Newcastle upon Tyne
Organisation for Economic Co-operation and Development [OECD] (2011) OECD territorial reviews: Slovenia 2011. OECD Publishing, Paris
Pavlinek P (2003) Alternative theoretical approaches to post-communist transformations in Central and Eastern Europe. Acta Slavica Iaponica 20:85–108
Power D, Nielsen T (2010) Priority sector report: creative and cultural industries. European Commision – Enterprise and Industry, Brussels
Pratt AC (2008) Creative cities: the cultural industries and the creative class. Geografiska Annaler Ser B Hum Geogr 90:107–117. doi:10.1111/j.1468-0467.2008.00281.x
Ravbar M (1997) Slovene cities and suburbs in transformation. Acta Geographica 37:65–109
Rekers JV (2012) We're number two! Beta cities and the cultural economy. Environ Plann A 44:1912–1929. doi:10.1068/a44511
Rutten R, Gelissen J (2008) Technology, talent, diversity and the wealth of European regions. Eur Plan Stud 16:985–1006. doi:10.1080/09654310802163785
Sassen S (1999) Digital networks and power. In: Featherstone M, Lash S (eds) Spaces of culture. Sage, London, pp 49–63
Scott AJ (1997) The cultural economy of cities. Int J Urban Reg Res 21:327–339. doi:10.1111/1468-2427.00075
Scott AJ (2001a) Capitalism, cities and the production of symbolic forms. Trans Inst Br Geogr 26:11–23. doi:10.1111/1475-5661.00003
Scott AJ (2001b) Globalization and the rise of city-regions. Eur Plan Stud 9:813–826. doi:10.1080/09654310120079788
Scott AJ (2006) Creative cities: conceptual issues and policy questions. J Urban Aff 28:1–17. doi:10.1111/j.0735-2166.2006.00256.x
Stryjakiewicz T, Męczyński M, Stachowiak K (2014) Role of creative industries in the post-socialist urban transformation. Quaestiones Geographicae 33:19–35. doi:10.2478/quageo-2014-0013
van Heerden S, Bontje M (2014) What about culture for the ordinary workforce? A study on the locational preferences of the creative class in Prenzlauer Berg, Berlin. J Urban Aff 36:465–481. doi:10.1111/juaf.12048

Chapter 12
Conclusion

Tadeusz Stryjakiewicz and Caroline Chapain

The contributions presented in this book reveal that there is still much to learn about the sectoral and spatial dynamics of the creative industries in Europe in social, economic and spatial research, despite the fast-growing interest they have aroused and the ever richer literature devoted to them in the last 15 years. This refers in particular to studies of these dynamics in a comparative approach, such as the one presented in this book. Several reasons explain this conclusion:

(a) While the creative industries have benefited from more than 15 years of research in Western Europe, their presence on the research agenda of Central and Eastern European countries has been shorter.
(b) Despite numerous reflections and studies examining the best way to define them and gathering evidence to document their sectoral and spatial dynamics, there are still debates with regard to their definition and classification due, among other things, to a substantial internal diversification of the cultural and creative sectors, gaps, and incompatibility in local, regional, national, and European statistics, and a great variety of research approaches and practical experiences across different geographical areas.
(c) The immaterial and symbolic characteristics associated with the creative industries present some uniqueness which make their comparison with other economic activities challenging.
(d) The necessity to take into account very different dynamics across geographical scales with, on the one hand, wide differences in the place that some creative

T. Stryjakiewicz (✉)
Institute of Socio-Economic Geography and Spatial Management, Adam Mickiewicz University, Poznań, Poland
e-mail: tadek@amu.edu.pl

C. Chapain
Business School, University of Birmingham, Birmingham, UK
e-mail: c.a.chapain@bham.ac.uk

© Springer International Publishing AG 2017

C. Chapain, T. Stryjakiewicz (eds.), *Creative Industries in Europe*,
DOI 10.1007/978-3-319-56497-5_12

industries occupy in national contexts and globally, on the other hand, their tendency to develop and be embedded in very specific types of local areas called creative milieux, creative fields, or creative clusters.

All those difficulties notwithstanding, the authors of all chapters have joined the discussion and tried to contribute a new 'value added' to the question posed in the title of the book, i.e. whether and in what circumstances creative industries are becoming real drivers of new development dynamics in the present European conditions and how can this inform the European debate about their contributions to both economy and society as discussed in the introduction. More particularly, questions arise about the detailed articulation of their sectoral and spatial dynamics within different national socio-economic and institutional contexts, especially in the context of Central and Eastern Europe. The analysis embraced two overlapping perspectives: sectoral and spatial. In addition, some chapters have more specifically focused on some remaining under-researched areas with regard to particular creative sector dynamics at the local levels, such as creative specialisation and diversification, creative clustering and innovation performance, creative concentration within a multi-scalar perspective and finally the role that cultural intermediaries can play in supporting the global positioning of both a creative industry and a city or the development of creative entrepreneurs at the local level.

The results presented in the first part of the book, entitled 'Interplay Between Sectoral and Spatial Dynamics of Creative Industries', confirm that 'interplay' is not merely a nice slogan. In consecutive chapters several forms of such an interplay were analysed. One of them is the interplay (or relationship) between the specialisation and diversification of creative cities and city regions discussed in the chapter by K. Stachowiak and T. Stryjakiewicz. On the basis of an overview of extensive literature and case studies, the authors proved that specialisation can be a driver of development for small- and medium-sized cities as well as large metropolises. But all of them use both specialisation and diversity to maintain their competitive advantages. That is why the authors propose to include 'diversified specialisation' to the typology of the specialisation of the European city regions (besides place-dependent and industry-dependent specialisation). They also confirm the results of earlier findings from M. Lorenzen and L. Frederiksen (2008, p. 176) who argue that 'urbanisation (regional diversity) may very well coexist with localisation (regional specialisation) and that many industries – among these the cultural industries – depend on this coexistence'. This coexistence embraces three interconnected dimensions: type of industry, labour market, and institutions.

In the geographical dimension, of particular importance is the interplay between the global and the local. While much space has been devoted to it in scholarly discourse, few detailed analyses can be found in the literature concerning location patterns of creative industries in a multi-scale approach, from the global to the district level. This gap is filled in the chapter by J. Molho and D. Sagot-Duvauroux, who use the case of art galleries. Their results indicate a growing polarisation of their location patterns, irrespective of the geographical region and scale of analysis, thus corroborating the general tendency towards clustering. At a global scale the art market is concentrated in three main macro-regions: Europe, North America and South-East Asia. In turn, in each of those regions there are cities (e.g. New York, London, Paris, Berlin, Zurich, Hong-Kong, or Shanghai) functioning as global art clusters. However, inter-urban patterns of concentration of galleries vary in different regions depending, on the one hand, on such 'classical' factors as wealth, urban and cultural assets, and connectivity, and on the other hand, on processes of urban land valuation and the collective effort of world art actors to be distinct from already established art districts. Most art galleries tend to be located in a few districts, either in affluent inner-city neighbourhoods or in semi-peripheral areas. The location dynamics of art galleries, as of most creative industries, is path-dependent.

In the next chapter, C. Chapain and K. Stachowiak, using the case of the film industry and the Soho cluster in London, discuss the interplay between innovation dynamics, geographical proximity, and the performance of this cluster. In their in-depth study, they showed the mechanisms of this interplay and confirmed the critical impact of geographical proximity and clustering to the innovative performance of the film industry in the UK (e.g. because of the high degree of embeddedness of this industry in the local labour market). However, international linkages within the film value chain as well as global competing film centres cannot be ignored, either. In particular, more research needs to be conducted on the role of local creative workers as integrator of both local and global knowledge and resources. This needs to better take into consideration local and global interplay echoes the analysis and findings from other chapters in this part and highlights the role of individuals as subtle nodes of connection in addition to institutions within this interplay as discussed in the chapter by D. Zajtmann. In search of other sources of new sectoral and spatial creative dynamics, D. Zajtmann in his chapter identifies and analyses one more type of interplay: that of a cultural intermediary like a professional organisation and the development of a creative industry (i.e. fashion) and a city within global networks. Here local and national professional organisations (meta-organisations), or 'cultural intermediaries', seem to play a key role, as illustrated by the fashion sector in Paris. Those organisations are also an important element of the interplay between creative industries and cities, and demonstrate – to use Scott's concept – that institutions of governance and collective action are a major component of the creative field of the city. This 'institutional approach' towards creative industries needs more elaboration in further research. In the case of Paris, the openness towards newcomers and international brands has been an important factor of the sustainability of the positon of the city in the high-end fashion industry globally.

The last type of interplay to be analysed in the last chapter of the first part of the book by B. Lange is one connected with the status of entrepreneurs in creative (or creative and cultural) industries. This is the interplay between their individual professionalisation and dependence on social contexts and professional scenes – another form of cultural intermediation. The author introduces an interesting analytical concept of entrepreneurial scenes and – using Berlin and Leipzig as examples – demonstrates that those scenes not only provide employment opportunities in creative and cultural industries, but they also structure entrepreneurial relation and professional networks (both formal and informal) and norms. Hence, particular local entrepreneurial scenes can be called yet another driver of the new sectoral and spatial dynamics of creative industries in Europe.

The focus of the contributions in the second part of the book, entitled 'National and Regional Spatial Dynamics of Creative Industries' shifts to the expression of creative sectoral dynamics at the regional and national scales. Hence, the predominant approach is spatial, acknowledging the importance of taking into account national economic geographies and socio-economic and institutional contexts when looking at the development of the creative industries. This part of the book confirms that 'geography matters', i.e. that modern Europe is a mosaic of different development pathways and that this mosaic embraces both the sectoral and spatial dynamics of creative industries.

A significant asset of this part of the book is that it presents the situation and dynamics of creative industries in various geographical regions of Europe: from Spain through Slovenia, the Czech Republic and Poland, to the youngest member of the European Union – Romania. Those are mostly countries in which the discourse on creative industries and their role as a development driver started relatively late, and knowledge about this matter is still rather skimpy. Therefore the presented results (including basic statistical data) fill a significant cognitive gap and give this discourse a truly European, comparative perspective.

Since the characteristics of the national and regional contexts of creative industries are discussed in individual chapters, below is an attempt to generalise the main features of their spatial dynamics in Europe in a comparative way.

1. The rise in the interest in and discourse on the creative industries and the path of their development is closely connected with the duration of a country's membership in the European Union. In the states of the 'old' Union (the case of Spain discussed in the chapter by M. Pareja-Eastaway and M. Pradel i Miquel), most economic indicators (including employment) show the share of creative industries to be higher than in the 'newcomers'. This finding, however, should be treated with great caution because of the incompatibility of data as mentioned above.
2. The spatial patterns of location of creative industries differ widely across Europe. In all European countries, a significant role is played by metropolitan areas, with the capital cities in the most prominent position. This confirms some of the findings of wider European studies as discussed in the introduction. However, while in countries with a polycentric settlement network (like Spain or Poland) there are also other large regional creative centres apart from the capitals, often competing with one another, in those with a monocentric network (like the Czech Republic or, to a lesser extent, Romania) the role of the capital is decidedly a dominant one. At the regional level, in the spatial patterns of creative industries, one can find suburbanisation effects. Those industries often tend to locate outside the core areas of cities, in suburban zones, with such excellent

examples here as Prague (see the chapter by O. Slach and J. Zenka) and Ljubljana (see the chapter by J. Kozina and D. Bole). This trend seems similar to that identified in many West European cities (see Bontje and Kepsu 2013).

3. The specificity of the sectoral and spatial dynamics of creative industries in the post-socialist countries of East-Central Europe can be expressed as follows:
   (a) These industries are included in global networks to a lesser extent than in Western Europe (with a very few exceptions, such as the gaming industry in Poland), but they play a significant role in the restructuring of local economies (mainly through newly established small- and medium-sized private firms).
   (b) In those countries, the role of clustering processes is smaller; they still seem to be at an initial stage and are poorly institutionalised, and emerging clusters are still developing their specialisation and branding.
   (c) Visible at the local scale – apart from a slight shift from the urban core towards the suburban hinterland – is the relationship between the location process of new creative firms and urban renewal processes (including the gentrification of degraded areas), strongly supported by the European Union.
4. One of the initial intentions of this book was to examine to what extent creative industries proved resistant to the 2008–2009 world financial crisis and helped to overcome its negative consequences in various European countries. However, this goal has only been accomplished in part. The conclusions concerning the resistance of creative industries to the crisis and their role in a post-crisis economic recovery are ambiguous. Some creative knowledge-intensive activities, such as computer programming, can undoubtedly be called new development drivers. At the opposite side stand many cultural industries. In our book their weak resistance to the crisis is evidenced by the cases of Romania (see the chapter by A. Becut and A. Craciun) and the Czech Republic (see the chapter by O. Slach and J. Zenka). The latter authors refer in their research on the impact of the crisis on the creative industries to three possible scenarios presented by A. Pratt (2009) and conclude that in the post-socialist countries a 'dependent scenario' seems predominant, as the development of creative industries in these countries has followed or even 'overtook' the development of the economy as a whole. Using the case of Czechia, they have not found any empirical proof that 'cultural production is driving the development of manufacturing, or the whole economy' (i.e. Pratt's third scenario). However, the situation differs even in neighbouring countries. For example, whereas in 2009–2013 Poland recorded growth or stabilisation in such sectors as publishing, film and TV or radio, and music production, in the Czech Republic their dynamics was much below the national average economy decline. In comparison with other post-socialist countries, in Poland, creative industries seem to have performed quite well during the crisis and do so also afterwards. Some of them (e.g. the gaming industry) are defined in official governmental documents as national specialisations with best prospects. Nevertheless, one can hardly generalise the main trends in the national and regional dynamics of creative industries because

their characteristic feature has been fluctuating from year to year. This is a topic which still demands further deeper studies based on long-term comparable data. Leaving aside statistics, it is worth mentioning the Spanish case (chapter by M. Pareja-Eastaway and M. Pradel i Miquel) which emphasises that, in addition, to a decline in economic output, local institutional setting was severely affected by the crisis, resulting in the weakening of some creative actors and the rise of new organisational forms within the creative sector.

5. The fluctuating trends in the sectoral and spatial dynamics of creative industries in Europe result at least partly from an absence of long-term integrated policies supporting their development. In many countries of Central and Eastern Europe, the European Commission's programme Creative Europe 2014–2020 has still not been transmitted to the regional and municipal levels. In our book one can find recommendations and examples of good practices aiming at new patterns of economic growth based on creative industries. In particular, worth mentioning are the experiences of Spanish cities presented by M. Pareja-Eastaway and M. Pradel i Miquel in their chapter. They include, among other things, policies of culture-led regeneration of city centres, new institutional and organisational patterns for creative industries (such as fab labs or co-working spaces), supporting the emergence of local creative scenes, strengthening connections of creative businesses with universities and research centres, creating spaces for dialogue and short-lasting relationships, 'liquid' companies, and, last but not the least, promoting a 'sophisticated' demand. More traditional policies, best illustrated by A. Becut and A. Craciun in their chapter on Romania, are being implemented in post-socialist countries of Eastern and Central Europe. They include professional training programmes (in order to ensure new skills for professionals working with new technologies, both in terms of production and distribution of cultural products), access to funds for small- and medium-sized creative enterprises, and incentives for public-private partnerships.

Worth mentioning and discussing further is also the IDEA model of determinants of the growth of the creative sector proposed by R. Kasprzak in his chapter. According to this model, four elements are crucial for making creative industries a real driver of new sectoral and spatial dynamics: internationalisation, demand, entrepreneurship, and availability. Although this model is very general, its application in Poland (as demonstrated by the author) seems quite successful.

Apart from the conclusions and insights arising from particular chapters, even more interesting and valuable are insights resulting from cross-referencing between chapters. These insights deepen the discussion about the development of the creative industries in Europe. Let us mention a few of them.

When discussing the issue of creative specialisation, K. Stachowiak and T. Stryjakiewicz focus to the spatial level of cities and city regions. However, as evidenced in the chapter by R. Kasprzak, in some countries (including those of Eastern and Central Europe) new national creative specialisations are emerging (such as the gaming industry in Poland). Hence, a question arises as to what extent national policies coincide with local policies in supporting the development of the creative sector. A comparison of Western and Eastern European experiences (e.g. the case of Spain analysed by M. Pareja-Eastaway and M. Pradel i Miquel vs. the case of Romania discussed by A. Becut and A. Craciun) suggests a greater divergence between the strategies and actions of local and national policy bodies in Eastern European countries. In those countries, local and national policies tend to follow very often different development approaches, e.g. the central government opts for the development of creative industries whereas the local authority prefers to focus on traditional industries (or vice versa, like in Poland recently).

In post-socialist countries, one more 'institutional clash' emerges. Indeed, the discussion around the development of the creative industries in the Czech Republic, Poland, Romania, and Slovenia seem to indicate that public support for creative industries results primarily from EU incentives (i.e. a top-down approach), and not so much from local governments and path-dependent conditions. A lack of 'tailored policies' and some deficiencies in public-private partnerships constitute another institutional obstacle for the development of the creative industries in this part of Europe. This is not the case of the former GDR, where – as described in the article by B. Lange – the attempts at integrating different actors' initiatives are more advanced.

In contrast, the chapters discussing the development of the creative industries in Western European cities and city regions, in particular, those described by C. Chapain and K. Stachowiak (London) and D. Zajtmann (Paris), show – contrary to the Central and Eastern European examples – that **local institutional contexts are crucial to understand the new patterns of growth of the creative sector**. This growth is usually (1) path-dependent, (2) clustered, and (3) strongly supported by institutions, professional organisations, and/or professional scenes. This institutional endowment facilitates global competition in such creative sectors as film or fashion. A similar statement can be applied to the development of art galleries discussed in detail by D. Sagot-Duvauroux and J. Molho. Apart from the features mentioned previously, they focus their analysis on development factors such as: connectivity and connective hubs, locational strategic differentiation, or real estate specificities. These factors seem neglected in the contributions presented by the authors from Eastern and Central Europe. In turn, the latter tend to emphasise the role of the creative industries as **drivers of post-socialist transformation** (both economic and non-economic). However, in their analyses of this transformation pathway, a parallel can be drawn with two of the Spanish modernisation pathways as distinguished by M. Pareja-Eastaway and M. Pradel i Miquel: one based (1) on a long-term planning process gathering consensus around some large urban/regional development projects and strategies to compete in the global arena or (2) on a more liberal approach, opting for short-term growth and taking advantage of national competitive niches. The majority of Eastern and Central European examples discussed in this book indicates a preference for the latter short-term pathway to support the development of the creative industries. This suggests a 'catching up' process in post-socialist European countries, but over the longer term, the Spanish case may be a good example to look at for these countries.

To summarise, it is difficult to say whether the creative sector constitute a universal way to support new economic development dynamics for all European regions. Standard quantitative evidence and economic indicators provided in the examples discussed in this book have not always confirmed such a direct leading role (its economic contribution seems indirect to a large extent, e.g. through its impact on innovation capacities). These findings may be explained in part by the low or high degree of development of the institutional settings and other creative intermediaries supporting these industries. While European documents, policies and initiatives have increasingly highlighted the role of the creative industries to support European competitiveness and some have focused on these issues recently, more research is needed to investigate which creative sectors and in what spatial and institutional settings can offer the highest added value not only in terms of economic aspects, but also non-economic ones such as social cohesion, a better quality of life, and the growth of other 'non-creative' sectors (so-called multiplier effects) across the various European regions. Moreover, this research will need to take into account the return towards a regional re-industrialisation advocated in some regions, i.e. the revitalisation of some traditional industries.

The contributions presented in this book prove that creative industries are undoubtedly one of the drivers of the present qualitative change of the European space and that beyond overall statistical analysis and evidence exercises conducted at the European level, it is important to foster a more in-depth understanding of the interplay of sectoral and spatial dynamics at the local, regional, and national levels in a comparative way.

References

Bontje M, Kepsu K (2013) Creative knowledge strategies for polycentric city-regions. In: Musterd S, Kovács Z (eds) Place-making and policies for competitive cities. Wiley-Blackwell, Oxford, pp 191–208

Lorenzen M, Frederiksen L (2008) Why do cultural industries cluster? Localisation, urbanization, products and projects. In: Cooke P, Lazzeretti L (eds) Creative cities, cultural clusters and local economic development. Edward Elgar, Cheltenham, pp 155–179

Pratt AC (2009) The creative and cultural economy and the recession. Geoforum 40(4):495–496

MIX
Papier aus verantwortungsvollen Quellen
Paper from responsible sources
FSC® C105338

If you have any concerns about our products,
you can contact us on
ProductSafety@springernature.com

In case Publisher is established outside the EU,
the EU authorized representative is:
Springer Nature Customer Service Center GmbH
Europaplatz 3, 69115 Heidelberg, Germany

Printed by Libri Plureos GmbH
in Hamburg, Germany